BASIC/NOT BORING
MATH SKILLS

COMPUTATION & NUMBERS

Grades 4–5

Inventive Exercises to Sharpen Skills and Raise Achievement

Series Concept & Development
by Imogene Forte & Marjorie Frank

Exercises by Sheri Preskenis

Illustrations by Kathleen Bullock

Incentive Publications, Inc.
Nashville, Tennessee

About the cover:

Bound resist, or tie dye, is the most ancient known method of fabric surface design. The brilliance of the basic tie dye design on this cover reflects the possibilities that emerge from the mastery of basic skills.

Cover art by Mary Patricia Deprez, dba Tye Dye Mary®
Cover design by Marta Drayton, Joe Shibley, and W. Paul Nance
Edited by Anna Quinn

ISBN 0-86530-403-3

PRINTED IN THE UNITED STATES OF AMERICA

TABLE OF CONTENTS

CELEBRATE BASIC MATH SKILLS

Basic does not mean boring! There is certainly nothing dull about . . .
. . . dropping in on an Olympic parade of athletes
. . . getting to know a whole list of Olympic records, feats, and trivia
. . . learning about some great sports tricks like Snatches, Wheelies, and McTwists
. . . following the sports drama of downhill racers and daring snowboarders
. . . finding out the difference between a sabre and a foil
. . . helping kayak racers get down a wild river without flipping
. . . figuring out who has won the most Olympic medals over the years

These are just a few of the interesting adventures students can explore as they celebrate basic math skills with computation and numbers. The idea of celebrating the basics is just what it sounds like—sharpening math skills while enjoying the investigation of real-life events. Each page of this book invites students to practice a high-interest math exercise built around facts and situations from the Summer and Winter Olympics. This is not just any ordinary fill-in-the-blanks way to learn. These exercises are fun and surprising, and they make good use of thinking skills. Students will do the useful work of practicing a specific math skill while stepping into the fascinating world of Olympic sports. They'll be delighted by facts about athletes, medals, venues, spectators, Olympic symbols, and great performances as they use Olympic statistics to examine numbers and solve problems.

The pages in this book can be used in many ways:

- to review or practice a math skill with one student
- to sharpen the skills with a small or large group
- to start off a lesson on a particular skill
- to assess how well a student has mastered a skill

Each page may be used to introduce a new skill, reinforce a skill, or assess a student's ability to perform a skill. And there's more than just the great student activities. You will also find an appendix of resources helpful to students and teachers—including a ready-to-use test for assessing computation and number skills.

As your students take on the challenges of these adventures with math, they will grow! And as you watch them check off the basic math skills they've strengthened, you can celebrate with them!

The Skills Test

Use the skills test beginning on page 56 as a pretest and/or a post-test. This test will help you check the students' mastery of basic computation and number skills and will prepare them for success on achievement tests.

SKILLS CHECKLIST FOR
COMPUTATION & NUMBERS, Grades 4-5

✔	SKILL	PAGE(S)
	Read and write whole numbers	10–12
	Solve word problems with whole numbers	10, 11, 19, 23, 27, 50, 51
	Compare and order whole numbers	13
	Round whole numbers	14
	Identify place value of whole numbers	15
	Add and subtract whole numbers	16–19, 50, 51
	Multiply whole numbers	19–22, 24–26, 50, 51
	Choose among a variety of whole number operations	19, 25, 50, 51
	Select the proper operation for a given computation	19, 27, 50, 51
	Identify common factors and greatest common factors	20, 21
	Solve equations with whole numbers	22–24, 26, 27, 50, 51
	Divide whole numbers	23–26, 50, 51
	Multiply and divide by powers of ten	24
	Solve multi-step problems with whole numbers	26
	Identify and use properties of operations	28
	Name fractional parts of a whole or set	29, 30
	Read and write fractional numbers and mixed numerals	29–41
	Solve word problems with fractional and decimal numerals	29, 30, 34, 40, 45, 46, 48
	Identify and write fractions in lowest terms	31
	Compare and order fractions	32, 33
	Identify equivalent fractions	34
	Write mixed numerals as fractions and fractions as mixed numerals	35, 36
	Write and solve equations with fractions and decimals	37–39
	Add and subtract fractions and mixed numerals	37–39
	Multiply and divide fractions and mixed numerals	40, 41
	Read and write decimals and mixed numerals	42–49
	Compare and order decimals and mixed numerals	43
	Round decimals	44
	Add and subtract decimals	45
	Solve problems with decimal numerals	45–49
	Multiply and divide decimals	46
	Change fractions to decimals and decimals to fractions	47–49
	Understand, read, and write percents	48, 49
	Write decimals as percents and percents as decimals	48, 49
	Write fractions as percents and percents as fractions	48, 49
	Read, write, compare, and order integers	52
	Add and subtract integers	52

COMPUTATION & NUMBERS

Skills Exercises

ATHLETES ON PARADE

The Olympic Games begin with a parade of all the athletes who will compete in the games. In the 1996 Summer Olympic Games in Atlanta, Georgia, 10,750 athletes came from 197 countries to compete for 17 days. Hundreds of medals were given out, millions of dollars were spent, thousands of visitors attended, and billions more watched on TV. In 1998, it started all over again when 3,000 athletes from 70 nations traveled to Nagano, Japan for the Winter Olympics.

Write the words that match the numbers for these Olympic facts.

1. There were 10,750 athletes competing in Atlanta.

 ____

2. The Atlanta Olympics cost 1,600,000,000 dollars.

 ____

3. About 2,000,000 people came to Atlanta to watch the 1996 Summer Olympics.

 ____

4. In 1896, there were about 40,000 spectators in Athens, Greece for the first of the modern Olympic Games. ____

5. There were over 3,000 hours of live TV coverage of Olympic events in Atlanta.

 ____

6. The television rights for the Barcelona Summer Games in 1992 were sold for 40,100,000 dollars. ____

7. At the Summer Olympics in Atlanta in 1996, 1,933 medals were handed out.

 ____

8. The Olympic Stadium in Atlanta held 85,000 spectators.

 ____

9. About 8,062 athletes have participated in the past four Winter Olympics (1988, 1992, 1994, 1998). ____

10. The city of Nagano, Japan has a population of 350,000.

 ____

Use with page 11.

Name ____

ATHLETES ON PARADE, CONT.

Write the numbers that match the words for these Olympic facts.

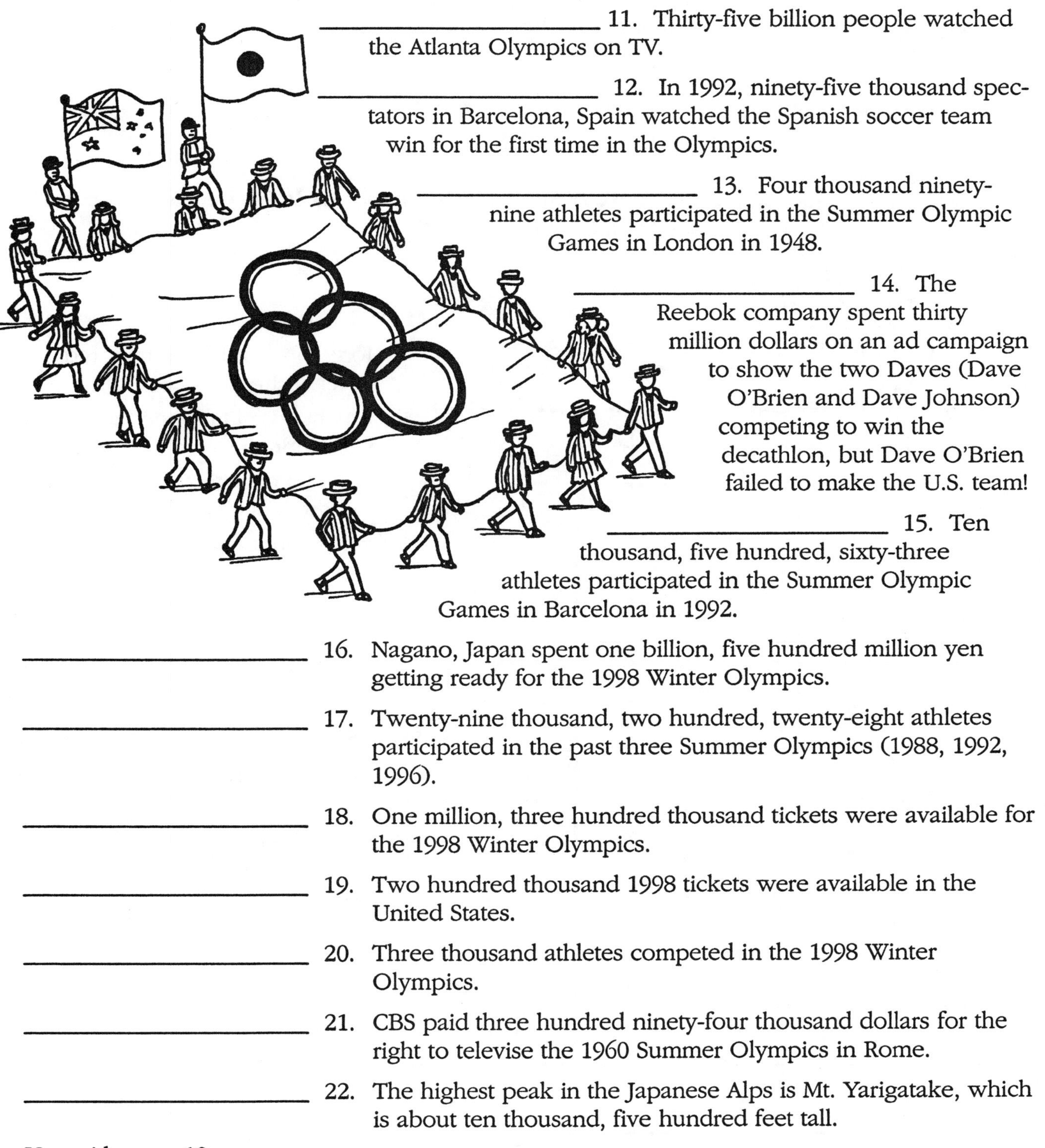

______________ 11. Thirty-five billion people watched the Atlanta Olympics on TV.

______________ 12. In 1992, ninety-five thousand spectators in Barcelona, Spain watched the Spanish soccer team win for the first time in the Olympics.

______________ 13. Four thousand ninety-nine athletes participated in the Summer Olympic Games in London in 1948.

______________ 14. The Reebok company spent thirty million dollars on an ad campaign to show the two Daves (Dave O'Brien and Dave Johnson) competing to win the decathlon, but Dave O'Brien failed to make the U.S. team!

______________ 15. Ten thousand, five hundred, sixty-three athletes participated in the Summer Olympic Games in Barcelona in 1992.

______________ 16. Nagano, Japan spent one billion, five hundred million yen getting ready for the 1998 Winter Olympics.

______________ 17. Twenty-nine thousand, two hundred, twenty-eight athletes participated in the past three Summer Olympics (1988, 1992, 1996).

______________ 18. One million, three hundred thousand tickets were available for the 1998 Winter Olympics.

______________ 19. Two hundred thousand 1998 tickets were available in the United States.

______________ 20. Three thousand athletes competed in the 1998 Winter Olympics.

______________ 21. CBS paid three hundred ninety-four thousand dollars for the right to televise the 1960 Summer Olympics in Rome.

______________ 22. The highest peak in the Japanese Alps is Mt. Yarigatake, which is about ten thousand, five hundred feet tall.

Use with page 10.

Name ______________

A BIG RACE—A WARM POOL

At the 1996 Summer Olympics, 14,000 people could find seats in the Aquatic Center to watch swimming events. Swimmers could compete in 32 different events for medals. They swam from 50 meters to 1500 meters in different races. The swimming pool was 50 meters long, with water that stayed at about 78° to 80° (Fahrenheit).

There are many numbers around the Olympic Games—numbers of people and medals, scores, distances, measurements, temperatures, and amounts of money.

Read these numbers written in words. Write the numerals into the puzzle to match the clues.

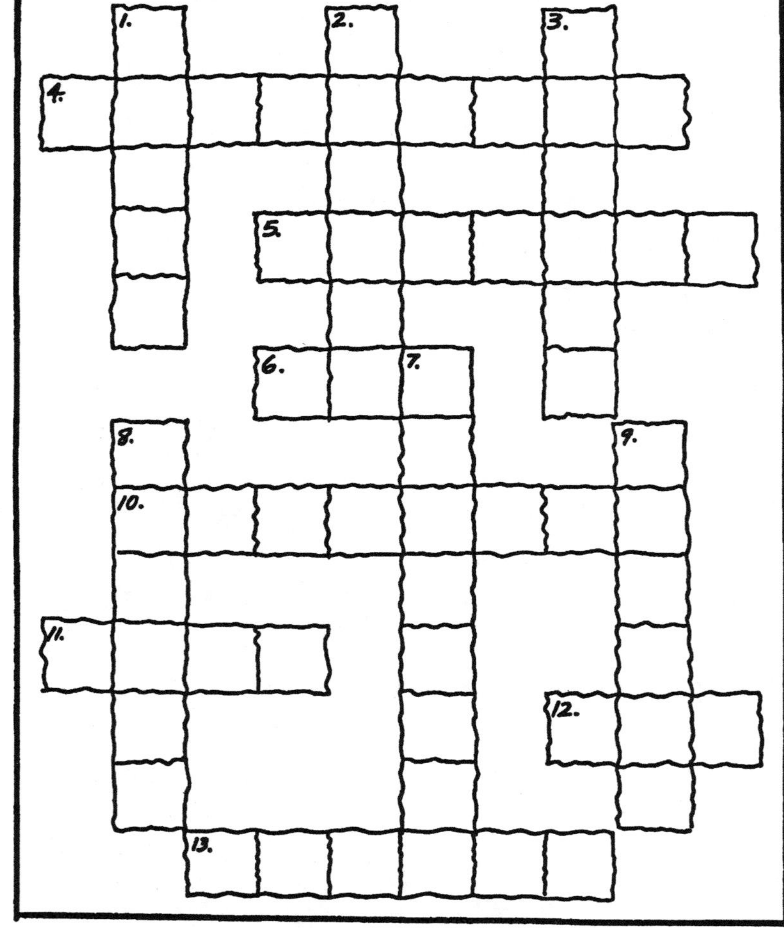

DOWN

1. forty thousand, nine hundred, seventy-three
2. one hundred fifty-one thousand, six hundred
3. nine hundred thousand, nine hundred, one
7. seventy-one million, eight hundred thousand, three
8. six hundred ten thousand, three hundred, ninety
9. four hundred fifty thousand, nine

ACROSS

4. three hundred million, fifty thousand, eight
5. two million, six hundred thousand, nine hundred
6. two hundred seven
10. twelve million, eight thousand, thirty-five
11. eight thousand, three hundred, fifty-one
12. nine hundred nine
13. three hundred thousand, three hundred

Name

SNATCH, CLEAN, & JERK!

Snatch, clean, and *jerk* may not sound like sports words—but they are! These strange words tell the names of the important moves in weight lifting. To perform a Snatch, the lifter brings the weight from a platform to an overhead hold in one movement. To perform a Clean & Jerk, the lifter pulls more weight, but the lift has two parts. First, the lifter brings the weight to his shoulders, and then he lifts it overhead.

1. Look at these weights lifted by some Olympic athletes. Read the numbers, and put them in order by numbering from 1 to 12 to show the numbers from smallest to largest, with 1 being the smallest.

_____ Alexandre 462 kg

_____ Leonid 425 kg

_____ Manfred 430 kg

_____ Martin 407 kg

_____ Dean 412 kg

_____ Mario 410 kg

_____ Sulton 440 kg

_____ Jurgen 411 kg

_____ Todeuz 408 kg

_____ Vassili 441 kg

_____ Helmut 387 kg

_____ Rudolf 610 kg

2. Number these from smallest to largest.

_____ 72,999

_____ 70,859

_____ 107,200

_____ 17,040

_____ 5,966

_____ 1,000,000

_____ 600,000

_____ 51,030

_____ 999,999

Olympic Fact

Tommy Kono was sick as a child. His parents tried to cure his asthma with powdered snakes, burned bird, and bear kidneys. During World War II, his family was sent to a Japanese-American detention camp. It was a terrible time for his family, but he was introduced to weight lifting there. Tommy won a gold medal as a weight lifter in 1952.

Name

Round Whole Numbers

THE ETERNAL FLAME

At the 1996 Summer Olympic Games, 10,750 athletes gathered under the Olympic flag to compete in 271 events from 26 different sports. There were 1,933 medals given. Many Olympic symbols were in view at the Opening Ceremonies. The most breathtaking symbol was the Olympic Torch. Its flame came from a torch that burns continuously in Athens, Greece. This "eternal flame" was originally ignited by the sun. Every Olympic year, the flame is carried to the host city by a torch relay. Many people take part in bringing the flame to the site of the latest Olympic Games.

When information about the Olympics is reported, the numbers are often rounded for easier repeating. Round these numbers to the digit that is underlined.

1. 10,750 ________________
2. 271 ________________
3. 1,933 ________________
4. 357 ________________
5. 891 ________________
6. 7,486 ________________
7. 15,426 ________________
8. 4,542 ________________
9. 800,426 ________________
10. 754,086 ________________
11. 999,999 ________________
12. 101,326 ________________
13. 60,600 ________________
14. 774,688 ________________
15. 10,987 ________________
16. 766 ________________
17. 98,922 ________________
18. 609 ________________
19. 555 ________________
20. 55,555 ________________
21. 923 ________________
22. 18,533 ________________
23. 600,001 ________________
24. 1,507 ________________

Name

CLEAN JUMPS, PLEASE!

Show jumping is the most thrilling and most televised of all the Olympic equestrian events. The rider takes the horse around a course full of obstacles at top speeds. On the way, the rider tries to keep control of the horse and to avoid trips, falls, or other faults!

Help this rider get around this course without errors. Write the place value for each underlined digit. Write the answer on each obstacle the horse must jump.

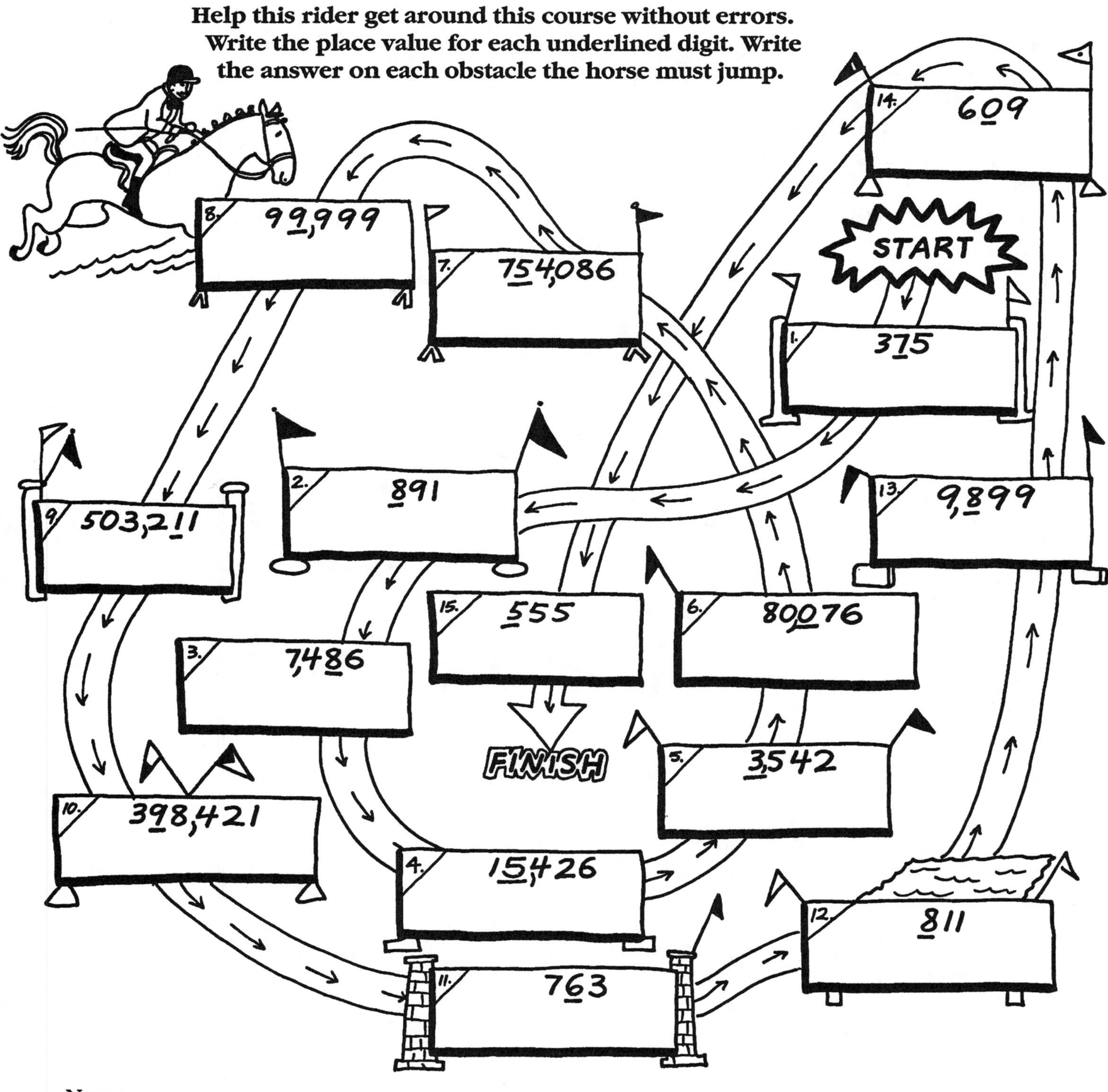

Name

HIGH SPEEDS & TOUGH TURNS

Experts say that the giant slalom takes the most technical skill of any ski event. Skiers race down the mountain over a long, steep, fast course. They must go through a series of gates marked by flags. Spectators also love to watch the downhill slalom, where skiers make high-speed turns to go through the gates at speeds of up to 80 miles per hour!

Write a long column addition problem using all the numbers on the slalom gate flags. Use box #1 to write and solve your problem. Then solve the other column addition problems.

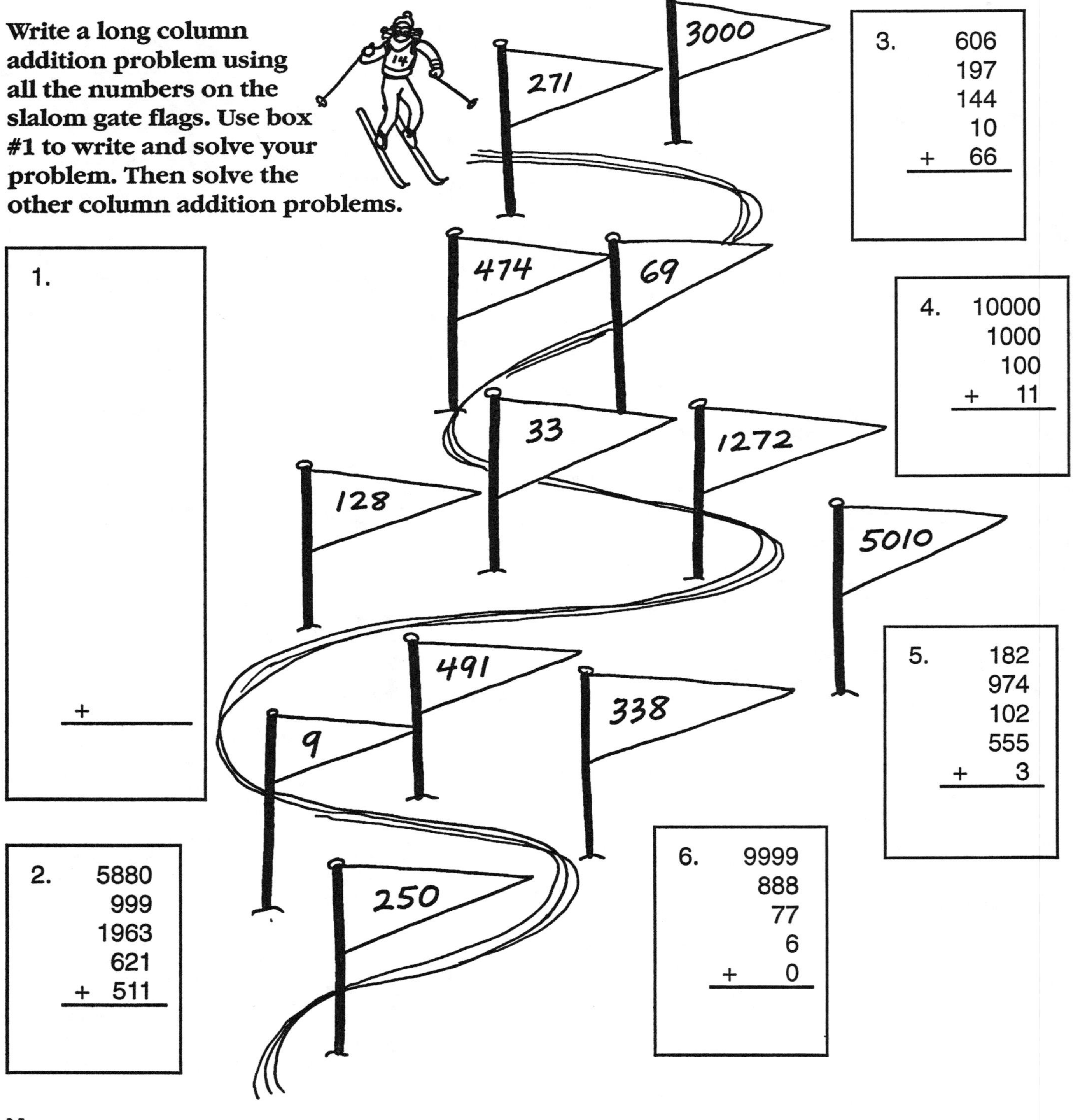

1.

+

2.
```
  5880
   999
  1963
   621
+  511
```

3.
```
   606
   197
   144
    10
+   66
```

4.
```
  10000
   1000
    100
+    11
```

5.
```
   182
   974
   102
   555
+    3
```

6.
```
  9999
   888
    77
     6
+    0
```

Name

KNOCK OUT!

Boxing was not allowed at the first modern Olympics in 1896 because it was considered too ungentlemanly and dangerous. Today it is a very popular Olympic sport. Some of the world's greatest boxers, such as Floyd Patterson, Muhammad Ali, Sugar Ray Leonard, Joe Frazier, Leon Spinks, and Evander Holyfield, won Olympic medals before becoming professional boxers.

See if you can knock out these subtraction problems by getting all the answers right!

1. 500 − 229 = ____

2. 900 − 683 = ____

3. 40 − 26 = ____

4. 300 − 258 = ____

5. 407 − 133 = ____

6. 90 − 55 = ____

7. 800 − 393 = ____

8. 5500 − 203 = ____

9. 9050 − 5348 = ____

10. 7001 − 6420 = ____

11. 6110 − 456 = ____

12. 8006 − 731 = ____

13. 800,321 − 79,001 = ____

14. 32,000 − 19,862 = ____

Olympic Fact

One of the most memorable moments of the 1996 Summer Olympic Games in Atlanta was when boxing legend and 1960 gold medal-winner Muhammad Ali (who suffers from Parkinson's disease) lit the Olympic torch.

Name ____________________

Add & Subtract Whole Numbers

EN GUARD!

Fencing was one of the events at the first Modern Olympic Games in 1896, but it began around 4000 B.C. Fencers use various types of swords: the foil, which weighs about 500 grams; the épée, which weighs 770 grams; and the sabre, which weighs 500 grams. When the director of the bout calls "en guard," the competitors take a ready position. They begin the "bout" when the director gives the "fence" command.

Try your skill in this bout with addition and subtraction.

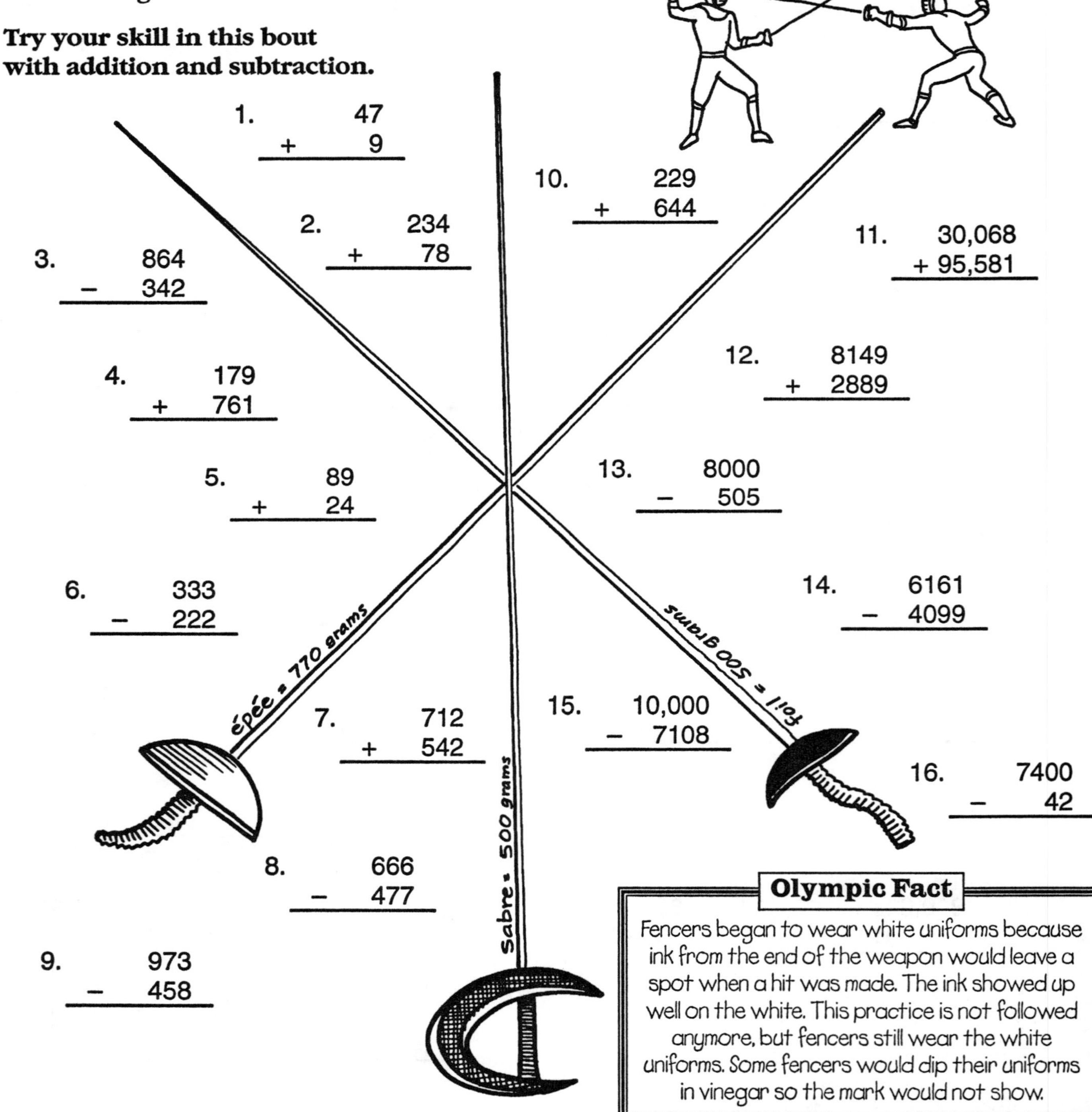

1. 47 + 9 = ____
2. 234 + 78 = ____
3. 864 − 342 = ____
4. 179 + 761 = ____
5. 89 + 24 = ____
6. 333 − 222 = ____
7. 712 + 542 = ____
8. 666 − 477 = ____
9. 973 − 458 = ____
10. 229 + 644 = ____
11. 30,068 + 95,581 = ____
12. 8149 + 2889 = ____
13. 8000 − 505 = ____
14. 6161 − 4099 = ____
15. 10,000 − 7108 = ____
16. 7400 − 42 = ____

Olympic Fact

Fencers began to wear white uniforms because ink from the end of the weapon would leave a spot when a hit was made. The ink showed up well on the white. This practice is not followed anymore, but fencers still wear the white uniforms. Some fencers would dip their uniforms in vinegar so the mark would not show.

Name ____________________

THE BIG WINNERS

TOP 30 MEDAL WINNERS
Olympic Summer Games
1896–1996

Country	Medals
United States	2,011
Soviet Union/ Russia	1,159
Germany	1,125
Great Britain	634
France	564
Sweden	461
Italy	445
Hungary	325
Australia	293
Finland	291
Japan	280
Romania	239
Poland	227
Canada	214
The Netherlands	188
Bulgaria	177
Switzerland	170
China	164
Czechoslovakia/ Czech Republic/ Slovakia	163
Denmark	134
Belgium	132
South Korea	126
Norway	121
Greece	111
Cuba	109
Yugoslavia	87
Austria	84
New Zealand	71
Spain	69
Turkey	59

The Modern Olympic Games began in 1896, 100 years before the Summer Games in Atlanta. Since then, thousands of medals have been given to hard-working athletes. The gold medal for first place is the most prized award! A silver medal is given for second place, and a bronze medal is given for third place. A new design for the medals is created for each Olympic Games. Solve the problems below with the information from the chart of top medal-winning countries.

1. Total medals won by the top 5 countries = __________
2. Germany's medals + Canada's = __________
3. ________________ won about 3 times as many as Cuba.
4. Great Britain won __________ fewer medals than the U.S.
5. This country won 211 fewer medals than Japan. ________
6. ________________ won about 4 times as many as Greece.
7. The Netherlands won ________ fewer medals than France.
8. Belgium and Denmark together won __________ medals.
9. Switzerland won __________ fewer medals than Germany.
10. The top 10 winners had a total of __________ medals.
11. Before the 1996 Olympics, Spain had a total of 46 medals. How many did Spain win in 1996? _________
12. Before the 1996 Olympics, the United States had a total of 1,910 medals. How many did the United States win in 1996? ________________________

Olympic Fact

The gold medal is not really made of gold. It is made mostly of silver, but it must contain at least six grams of pure gold.

Name

EXPLOSIVE SPEEDS

Olympic Fact

It never snows in Jamaica. The average temperature is 80°! Even so, Jamaica has a bobsled team. In the 1994 Winter Olympics, Jamaica's team finished 14th—ahead of both U.S. teams!

People began sledding over 15,000 years ago on sleds made of a strip of animal skin stretched between two pieces of wood. These days, Olympic bobsleds are high-tech machines made for speed. Bobsledding is thrilling to watch! A crew of two or four flies down a mile-long, curvy course at speeds of up to 90 mph. An explosive start is very important for a fast racing time.

See how fast you can find factors for these bobsleds. Write four factors for the number on the sled. Write one factor on the helmet of each crew member.

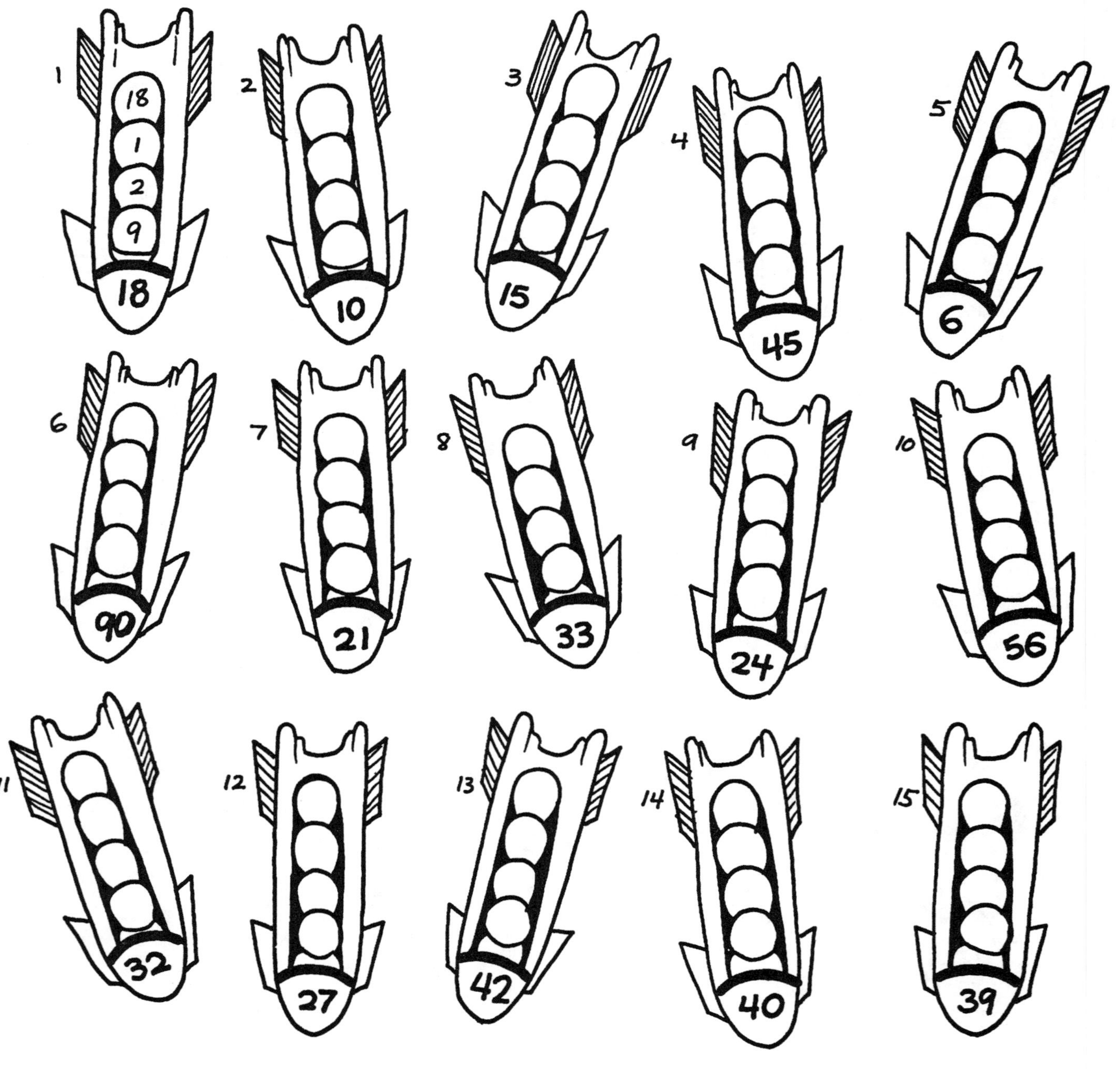

Name

NO BRAKES!

It is said that some of the earliest bobsleds had no brakes and were steered by a rope! The sleds were stopped in a very interesting way. Find out how by solving the puzzle below.

Write the greatest common factor for each pair of numbers. Then write the corresponding letter to your answer on the line above that answer in the puzzle at the bottom of the page. If your answers are correct, you will find out how the sleds were stopped.

1. 2 and 4 _____ R
2. 9 and 6 _____ E
3. 11 and 33 _____ T
4. 14 and 28 _____ I
5. 30 and 45 _____ V
6. 18 and 24 _____ N
7. 28 and 32 _____ D
8. 15 and 27 _____ E
9. 80 and 50 _____ A
10. 12 and 36 _____ K
11. 3 and 9 _____ E
12. 4 and 12 _____ D
13. 6 and 10 _____ R
14. 16 and 6 _____ R
15. 14 and 10 _____ R
16. 10 and 40 _____ A
17. 5 and 25 _____ G
18. 15 and 50 _____ G
19. 16 and 4 _____ D
20. 7 and 35 _____ H
21. 12 and 32 _____ D
22. 40 and 90 _____ A
23. 5 and 15 _____ G
24. 20 and 10 _____ A
25. 22 and 6 _____ R
26. 15 and 9 _____ E
27. 21 and 3 _____ E

__ __ __   __ __ __ __ __ __   __ __ __ __ __ __ __
11 7 3   4 2 14 15 3 2   4 2 10 5 5 3 4

__   __ __ __ __ __ __   __ __ __ __ to stop the sled.
10   5 10 2 4 3 6   2 10 12 3

Name

MAY THE BEST SAILOR WIN

Yachting has been an Olympic sport since the 1896 games in Athens. Unfortunately, the yachting races had to be canceled at those games! The weather was just too bad. In each racing class, all the yachts must have the same design. This way, the best sailor wins the race, not the best boat.

Solve the multiplication problems in the puzzle. Use the color code to find the color for each section. If you get the answers right, the colored picture will show you one kind of yacht used in Olympic racing.

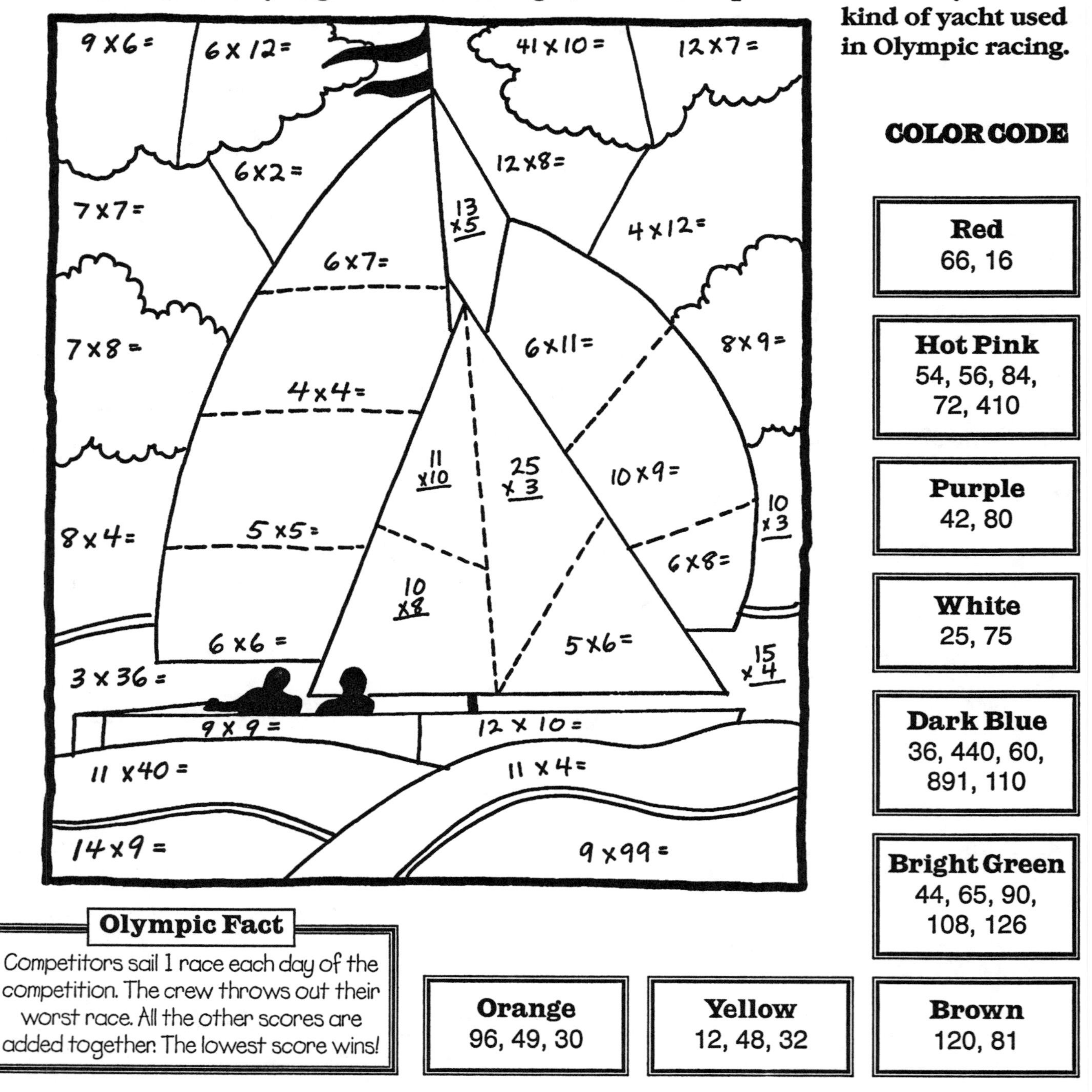

COLOR CODE

Red	66, 16
Hot Pink	54, 56, 84, 72, 410
Purple	42, 80
White	25, 75
Dark Blue	36, 440, 60, 891, 110
Bright Green	44, 65, 90, 108, 126
Orange	96, 49, 30
Yellow	12, 48, 32
Brown	120, 81

Olympic Fact

Competitors sail 1 race each day of the competition. The crew throws out their worst race. All the other scores are added together. The lowest score wins!

Name

FANS BY THE THOUSANDS

When a country hosts the Olympic Games, they spend many months and a lot of money getting ready. Most countries try to use sports arenas and areas that they already have, but many new buildings and venues must be built for all the events and the spectators. Usually a country builds a new Olympic Stadium. The stadium in Atlanta was built to hold 85,000 fans.

If the 85,000 seats in Atlanta were arranged in 50 equal sections, how many seats would there be in each section? To find the answer, you would need to divide 85,000 by 50. Use division to find the answers to these problems.

1. Aquatic Center—swim events 14,000 seats ÷ 20 sections = ___________
2. Georgia World Congress—fencing, judo 7,500 seats ÷ 25 sections = ___________
3. Georgia Tech Coliseum—boxing 9,500 seats ÷ 10 sections = ___________
4. Nagano's Hockey Arena—hockey 10,000 seats ÷ 50 sections = ___________
5. White Ring—speed skating 7,300 seats ÷ 5 sections = ___________
6. Nagano Olympic Stadium 50,000 seats ÷ 50 sections = ___________
7. Atlanta Olympic Stadium 85,000 seats ÷ 50 sections = ___________
8. Clark University Stadium—field hockey 5,000 seats ÷ 25 sections = ___________
9. Georgia Dome—basketball 32,000 seats ÷ 8 sections = ___________
10. Omni Coliseum—baseball 52,000 seats ÷ 40 sections = ___________

11. $9\overline{)8190}$ 12. $3\overline{)1227}$ 13. $6\overline{)9534}$ 14. $8\overline{)46,328}$

Olympic Fact

Atlanta spent $500 million on new buildings for the 1996 Olympics. The Olympic Stadium cost $209 million.

Name ____________________

SAILING WITHOUT A SAIL

It's spectacular . . . breathtaking . . . awesome! Crowds at the Winter Olympics always love to watch the ski jumpers sailing through the air. Spectators hold their breath until the skier lands safely on the ground! Skiers gain points for strong take-offs, smooth flights, clean landings, and distance. Skiers take off into the air from jumps as high as 120 meters and sail for hundreds of feet.

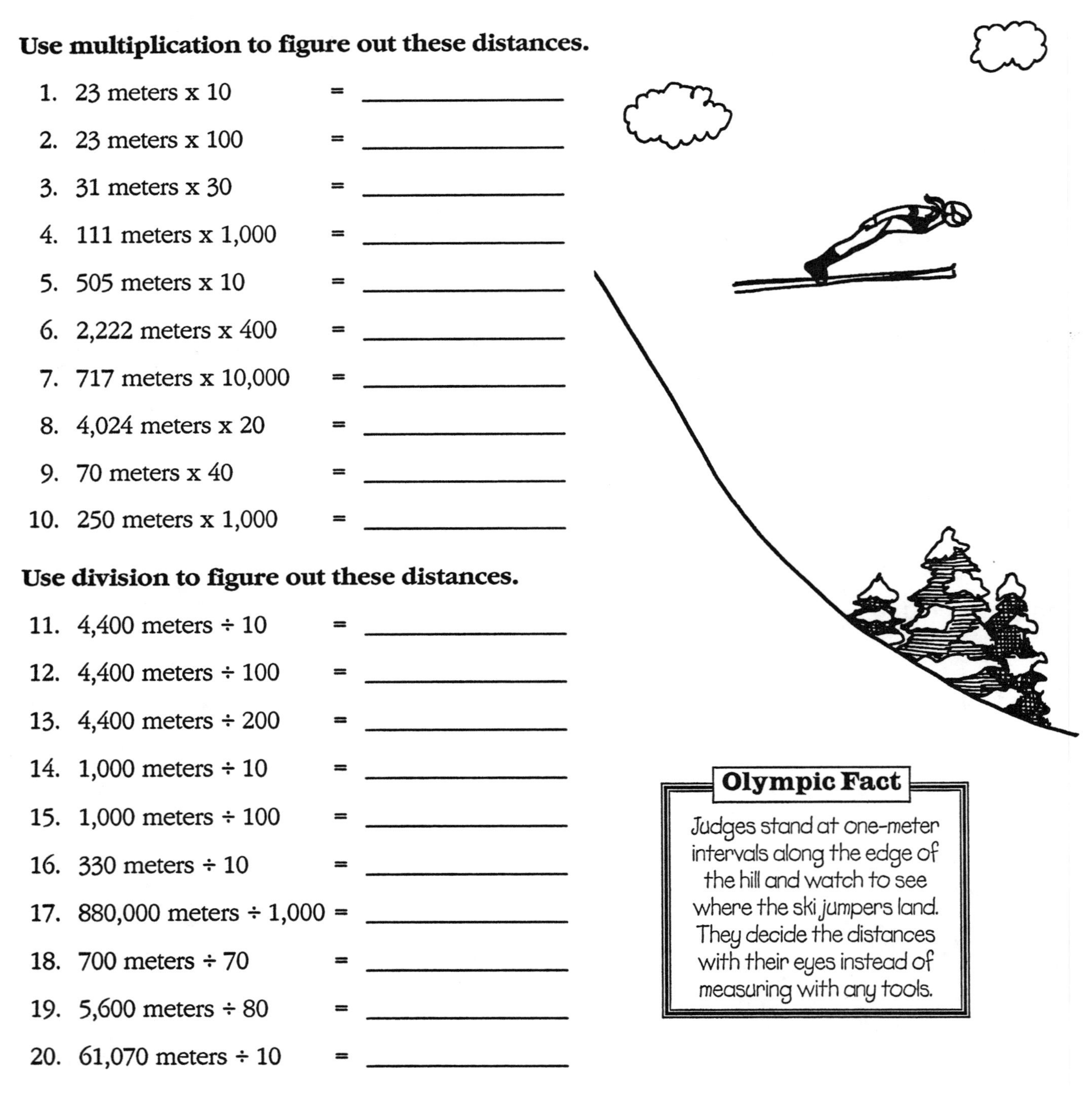

Use multiplication to figure out these distances.

1. 23 meters x 10 = ______________
2. 23 meters x 100 = ______________
3. 31 meters x 30 = ______________
4. 111 meters x 1,000 = ______________
5. 505 meters x 10 = ______________
6. 2,222 meters x 400 = ______________
7. 717 meters x 10,000 = ______________
8. 4,024 meters x 20 = ______________
9. 70 meters x 40 = ______________
10. 250 meters x 1,000 = ______________

Use division to figure out these distances.

11. 4,400 meters ÷ 10 = ______________
12. 4,400 meters ÷ 100 = ______________
13. 4,400 meters ÷ 200 = ______________
14. 1,000 meters ÷ 10 = ______________
15. 1,000 meters ÷ 100 = ______________
16. 330 meters ÷ 10 = ______________
17. 880,000 meters ÷ 1,000 = ______________
18. 700 meters ÷ 70 = ______________
19. 5,600 meters ÷ 80 = ______________
20. 61,070 meters ÷ 10 = ______________

Olympic Fact

Judges stand at one-meter intervals along the edge of the hill and watch to see where the ski jumpers land. They decide the distances with their eyes instead of measuring with any tools.

Name

HOLD YOUR BREATH!

Can you imagine doing a complicated routine underwater, in perfect timing with other athletes for $3\frac{1}{2}$ minutes, while holding your breath for most of the time? Synchronized swimmers do just that. In this Olympic sport, swimmers spend 60% of the routine time underwater. They must do difficult, synchronized movements, and touching the bottom or side of the pool is not allowed!

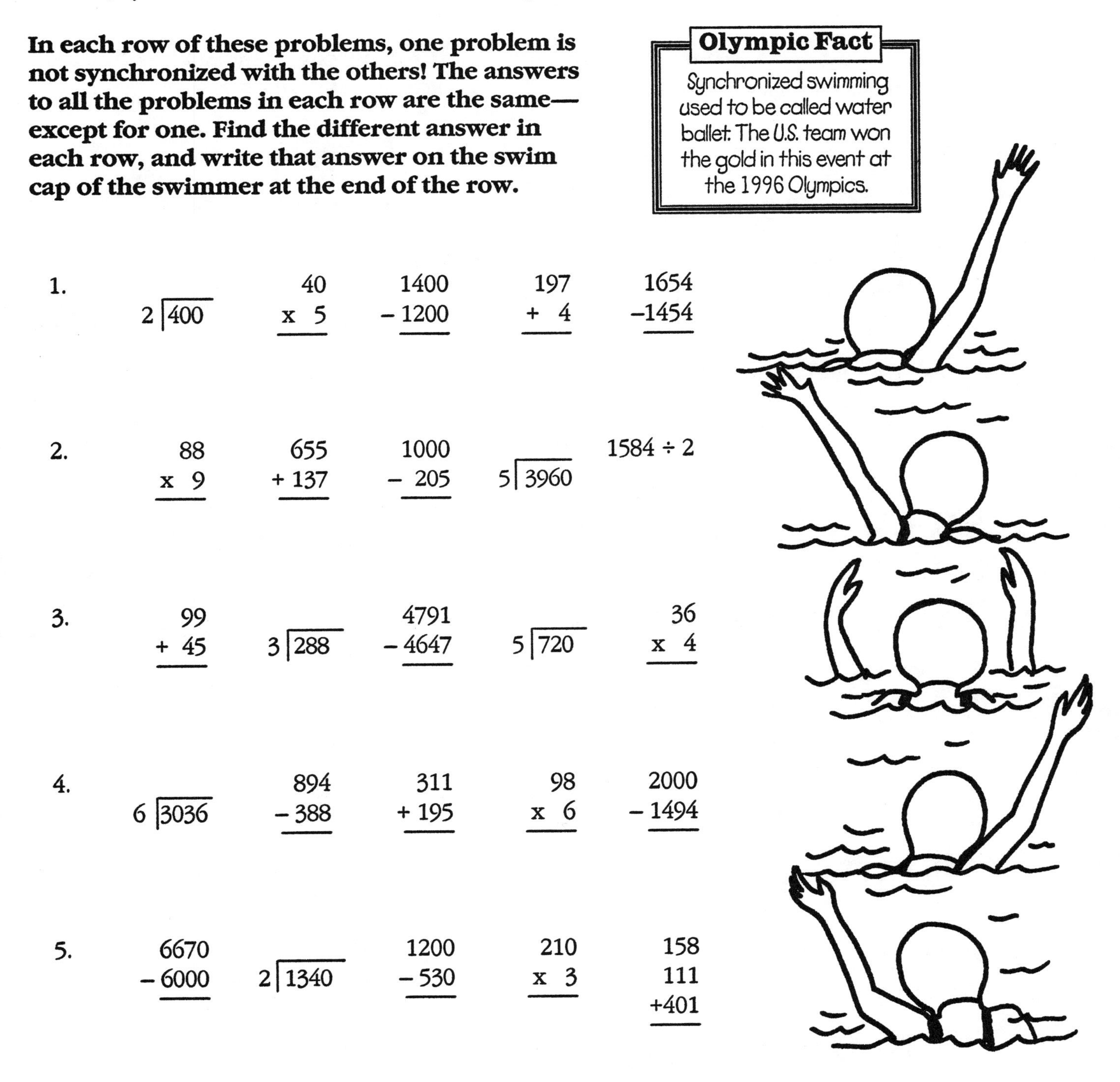

In each row of these problems, one problem is not synchronized with the others! The answers to all the problems in each row are the same—except for one. Find the different answer in each row, and write that answer on the swim cap of the swimmer at the end of the row.

Olympic Fact

Synchronized swimming used to be called water ballet. The U.S. team won the gold in this event at the 1996 Olympics.

1.	$2\overline{)400}$	40×5	$1400 - 1200$	$197 + 4$	$1654 - 1454$
2.	88×9	$655 + 137$	$1000 - 205$	$5\overline{)3960}$	$1584 \div 2$
3.	$99 + 45$	$3\overline{)288}$	$4791 - 4647$	$5\overline{)720}$	36×4
4.	$6\overline{)3036}$	$894 - 388$	$311 + 195$	98×6	$2000 - 1494$
5.	$6670 - 6000$	$2\overline{)1340}$	$1200 - 530$	210×3	$158 + 111 + 401$

Name

MAKING IT OVER HURDLES

The Olympic hurdle event is a fast sprinting race with a series of barriers to jump. The hurdles are made of wood and metal, and sometimes runners knock them over as they jump. It doesn't disqualify a hurdler to knock one over, but usually it slows him or her down a little. Men and women compete in hurdle events of different lengths and with different height hurdles. (Women: 2 ft 6 in. and 2 ft 9 in.; Men: 3 ft and 3 ft 6 in.)

Help this runner clear her hurdles by solving the problems correctly. Start with the first number, and do all the operations shown to find the final answer.

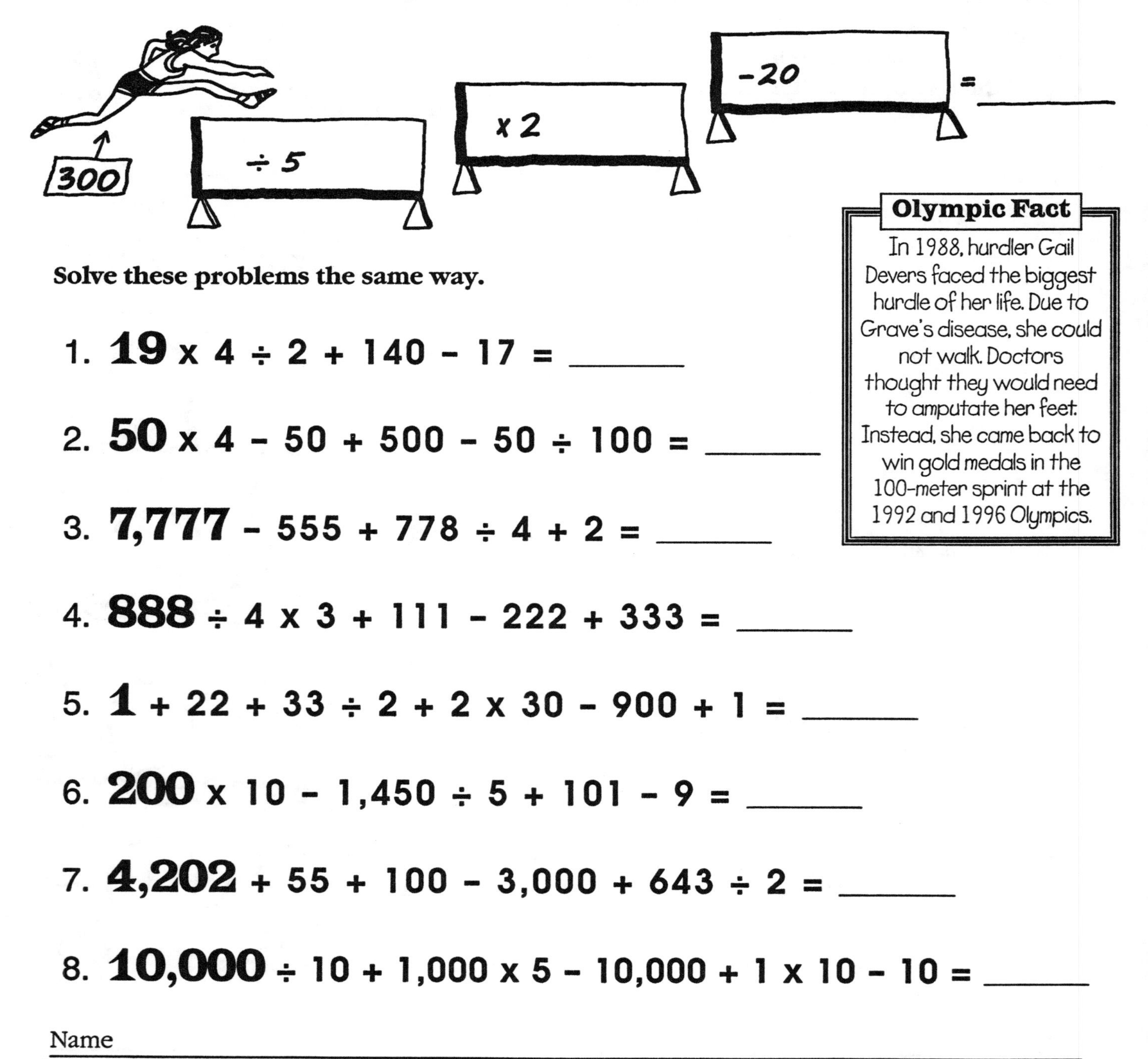

Solve these problems the same way.

1. **19** x 4 ÷ 2 + 140 − 17 = ______
2. **50** x 4 − 50 + 500 − 50 ÷ 100 = ______
3. **7,777** − 555 + 778 ÷ 4 + 2 = ______
4. **888** ÷ 4 x 3 + 111 − 222 + 333 = ______
5. **1** + 22 + 33 ÷ 2 + 2 x 30 − 900 + 1 = ______
6. **200** x 10 − 1,450 ÷ 5 + 101 − 9 = ______
7. **4,202** + 55 + 100 − 3,000 + 643 ÷ 2 = ______
8. **10,000** ÷ 10 + 1,000 x 5 − 10,000 + 1 x 10 − 10 = ______

Olympic Fact

In 1988, hurdler Gail Devers faced the biggest hurdle of her life. Due to Grave's disease, she could not walk. Doctors thought they would need to amputate her feet. Instead, she came back to win gold medals in the 100-meter sprint at the 1992 and 1996 Olympics.

Name

TOURIST ATTRACTIONS

Two million people attended the 1996 Olympics. They bought tickets; watched events; traveled to different venues; toured Atlanta; lived in hotels, tents, campers, and homes; and bought a lot of food and souvenirs.

Decide which operation you should use to solve each of these problems about Olympic tourists. Write the operation (add, subtract, multiply, or divide) after each problem. Then solve the problem.

1. Tickets for the kayaking race cost $27. The ticket office counted $29,403 for this event. How many tickets were sold?

   Operation ________________ **Answer** ____________

2. If hot dogs sold for $2.00 in the Olympic Park and 986,443 hot dogs were sold, how much money was collected?

   Operation ________________ **Answer** ____________

3. 26,000 bus and limo drivers were hired for the Olympics. If $\frac{3}{5}$ worked every day, how many worked at one time?

   Operation ________________ **Answer** ____________

4. One family drove 243.33 km from their home to Atlanta for the games. They went home by a route 21.7 km longer. How long was their trip home?

   Operation ________________ **Answer** ____________

5. In Nagano, 10,000 people could attend a hockey game at once. Of these, $\frac{1}{5}$ had "standing room only" tickets. How many fans had to stand?

   Operation ________________ **Answer** ____________

6. Brielle's family bought 18 Olympic basketballs as souvenirs to take home to friends. They spent $412.20. How much did each ball cost?

   Operation ________________ **Answer** ____________

7. Not all sports fans can get to the Olympics, so they are televised around the world. TV rights cost $2,500,000 in 1968. In 1992, they cost 120 times that much. How much did the 1992 rights cost?

   Operation ________________ **Answer** ____________

8. In Nagano, tickets for good seats at the Opening Ceremony cost $350. In Atlanta, the tickets cost $636. How much did Joanna's family of 4 pay to go to both?

   Operations ________________ & ________________

   Answer ____________

Name

THE RIGHT PROPERTIES

Hockey players have some special properties—the equipment they need to wear on the ice. Without the right stuff, they wouldn't be able to play the game very well (or very safely)! Their equipment certainly helps them with problems on the ice.

Math operations have properties that you need to use for solving math problems. Review these properties. Then decide which ones are used in the problems below. Write the name of the property below each problem.

Zero Property of Addition
Zero Property of Subtraction
Zero Property of Multiplication
Property of One
Opposites Property of Addition
Commutative Property of Addition
Commutative Property of Multiplication

1. 8 x 4 = 32 and 4 x 8 = 32

____

2. 8633 x 1 = 8633

____

3. 99 x 0 = 0

____

4. 110 x 55 = 6050 and 55 x 110 = 6050

____

5. 1 x 99 = 99

____

6. 6 x 8 = 8 x 6

____

7. 53 + 17 = 17 + 53

____

8. 666 – 0 = 666

____

9. 25 + 10 = 35 and 35 – 10 = 25

____

10. 1700 + 0 = 1700

____

11. 7401 x 15 = 15 x 7401

____

12. 77 x 0 = 0

____

Name _______________

WATCH THAT PUCK!

These fans are gathered for an exciting, high-speed ice hockey game. All the action in the game is focused on a little rubber disc that moves so fast that often it is hard to tell where it is and which team has it! An exciting Olympic moment for the United States was in 1980 when the U.S. team defeated Finland to win its first gold medal in 20 years.

Olympic Fact

The 1998 Winter Olympics in Japan were the first Games that permitted women to compete in ice hockey.

Pay attention to these fans to practice your fraction-hunting skills. Write a fraction to fill each blank.

1. ______ of the fans are holding balloons.
2. ______ of the fans are holding flags.
3. ______ of the flags have words on them.
4. ______ of the flags are black.
5. ______ of the flags have no words.
6. ______ of the fans are holding cups.
7. ______ of the cups have 2 straws.
8. ______ of the cups have no straws.
9. ______ of the fans are wearing boots.
10. ______ of the shoes and boots have black on them.
11. ______ of the fans are wearing earmuffs.
12. ______ of the fans are wearing hats.
13. ______ of the shoes and boots have laces.
14. ______ of the hands are wearing mittens or gloves.
15. ______ of the fans are wearing scarves.
16. ______ of the fans are hatless.
17. ______ of the hats have feathers.
18. ______ of the fans have mustaches.
19. ______ of the balloons are held by the girl with pigtails.

Name

FROSTY SPORTS

Most sports at the Winter Olympics are outdoor sports. Even if the competitions are held indoors, the temperatures are usually cold to keep the ice from melting. You'll need to be good with fractions to solve these puzzles with facts about the Winter Olympics.

QUESTION: *Which event has brought the most Winter Olympic medals to the U.S.?*

1. Write the second $\frac{1}{6}$ of SKIING. **K**
2. Write the first $\frac{1}{4}$ of ATTITUDE. ____________
3. Write the first $\frac{1}{3}$ of SPECTATOR. ____________
4. Write the first $\frac{1}{4}$ of GOLD. ____________
5. Write the last $\frac{1}{5}$ of SNOWBOARDS. ____________
6. Write the first $\frac{1}{10}$ of ICE DANCING. ____________
7. Write the last $\frac{1}{4}$ of GAME. ____________
8. Write the last $\frac{1}{8}$ of BIATHLON. ____________

ANSWER: *Unscramble the letters to find the sport of* ______________________________

QUESTION: *Which ski event brought U.S. skier Picabo Street a silver medal in 1992?*

1. Write the first $\frac{1}{6}$ of NAGANO. ____________
2. Write the first $\frac{1}{4}$ of LUGE. ____________
3. Write the second $\frac{1}{5}$ of SNOWBOARDS. ____________
4. Write the second $\frac{1}{8}$ of OLYMPICS. ____________
5. Write the third $\frac{1}{5}$ of MEDAL. ____________
6. Write the first $\frac{1}{5}$ of HIGHLIGHTS. ____________

ANSWER: *Unscramble the letters to find this event:*

____

Olympic gold medal-winner Dan Jansen has skated over 100,000 miles on his speed skates. This is more than 4 times the distance around the world.

Name ______________________________

BE CAREFUL NOT TO SWING!

Some of the most difficult moves a male gymnast must do are done while he hangs from rings. Gymnasts show amazing skill and strength as they hold their bodies in tough positions. The rings are not supposed to swing or wobble as the gymnast does the moves! The gymnast's body and arms are not supposed to wobble, sag, or shake!

Keep these rings from wobbling by identifying the ones with lowest term fractions. If a ring holds a fraction in lowest terms, color the inside of the ring with a marker or colored pencil. If the fraction is not in lowest terms, write the lowest term fraction in the ring.

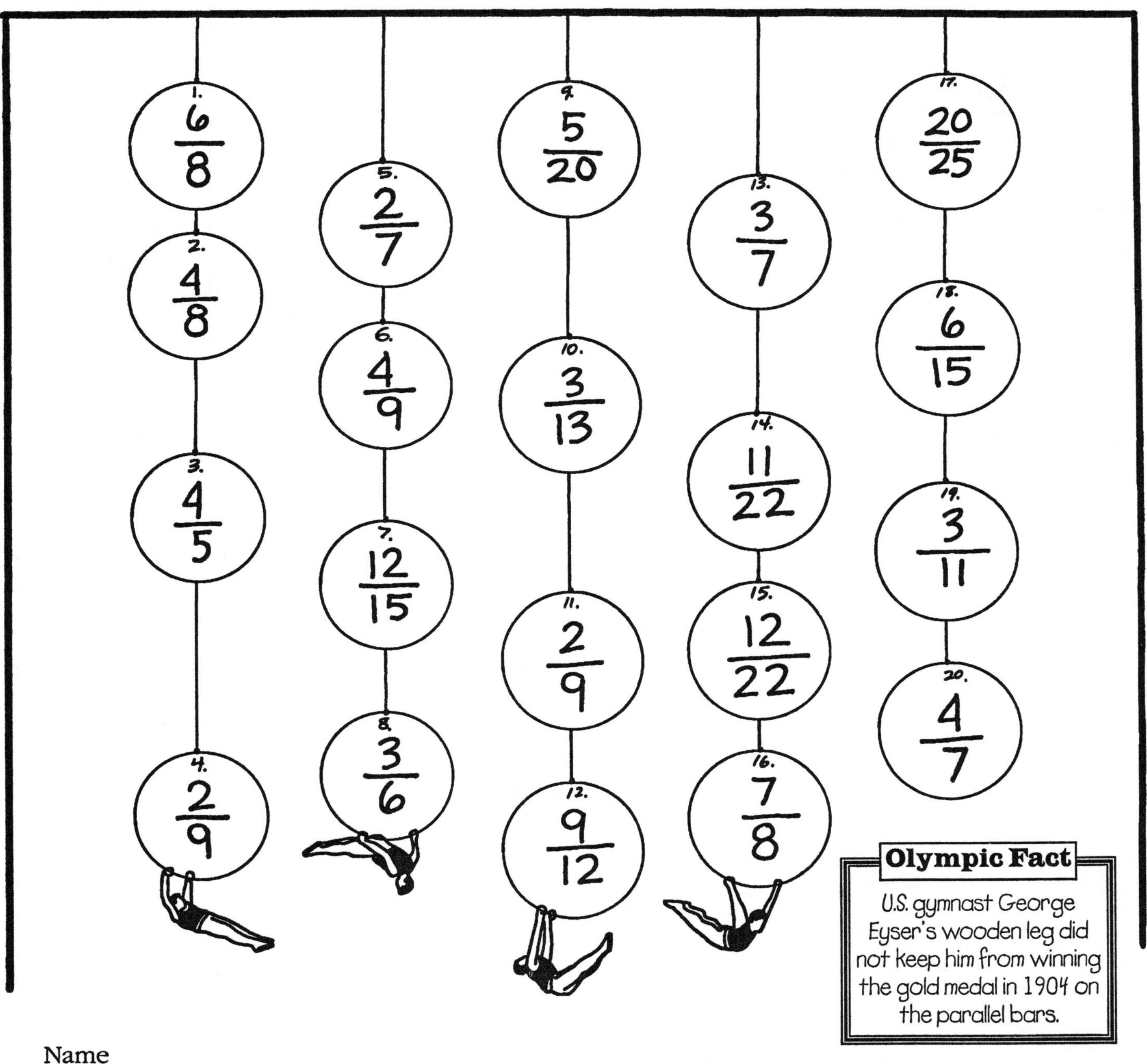

Olympic Fact

U.S. gymnast George Eyser's wooden leg did not keep him from winning the gold medal in 1904 on the parallel bars.

Name

OVER THE NET

> **Olympic Fact**
> In beach volleyball, each team has only two players. They play barefoot in the sand.

Beach volleyball began in the 1940s on the beaches of California. It was played for fun at first, but now it is a serious professional sport. It did not gain a place at the Olympic Games until 1996, when the U.S. men's teams won the gold and silver medals.

Compare each set of fractions below to see which is greater. Circle the largest fraction. If the fractions are equal, circle them both!

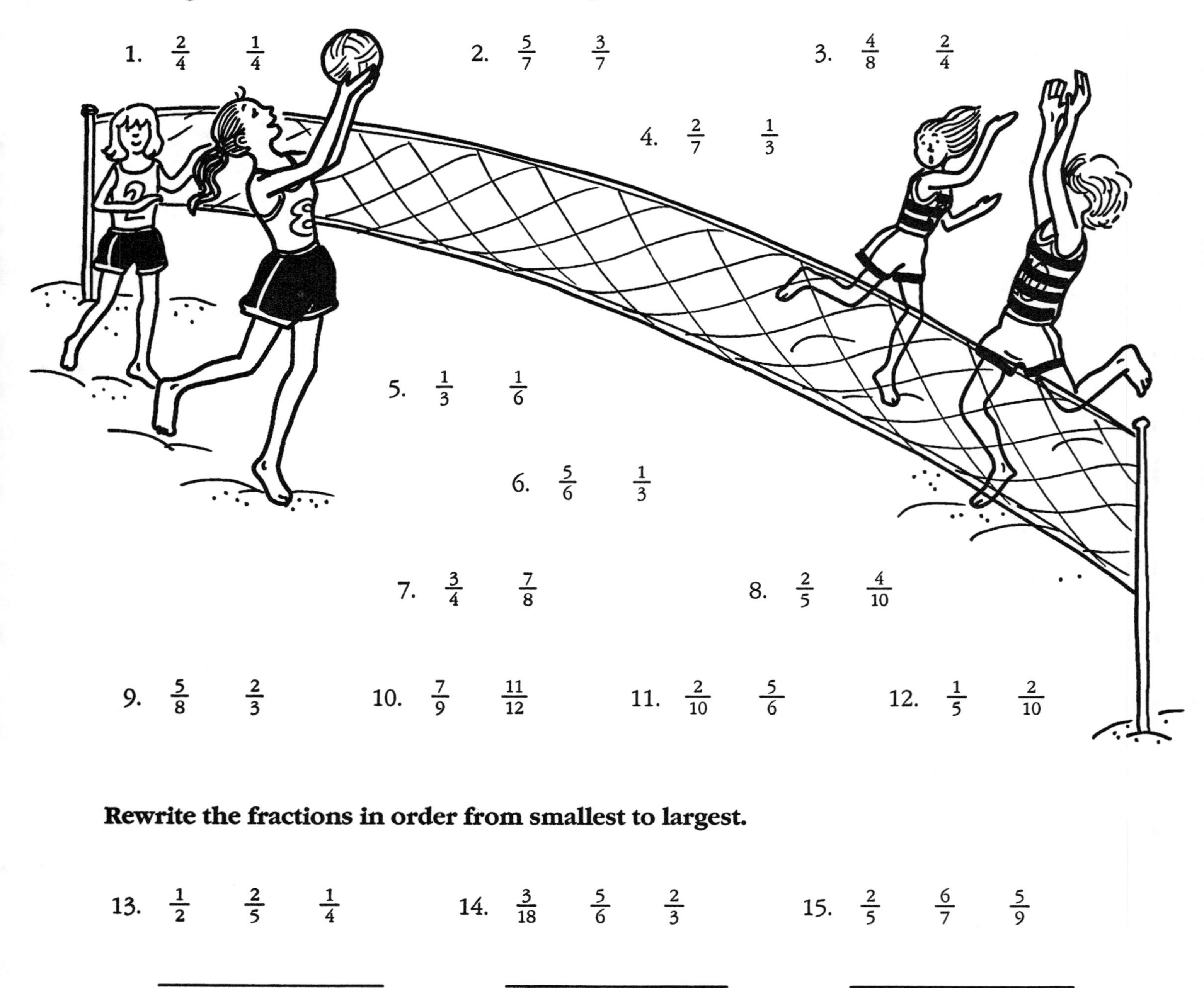

1. $\frac{2}{4}$ $\frac{1}{4}$
2. $\frac{5}{7}$ $\frac{3}{7}$
3. $\frac{4}{8}$ $\frac{2}{4}$
4. $\frac{2}{7}$ $\frac{1}{3}$
5. $\frac{1}{3}$ $\frac{1}{6}$
6. $\frac{5}{6}$ $\frac{1}{3}$
7. $\frac{3}{4}$ $\frac{7}{8}$
8. $\frac{2}{5}$ $\frac{4}{10}$
9. $\frac{5}{8}$ $\frac{2}{3}$
10. $\frac{7}{9}$ $\frac{11}{12}$
11. $\frac{2}{10}$ $\frac{5}{6}$
12. $\frac{1}{5}$ $\frac{2}{10}$

Rewrite the fractions in order from smallest to largest.

13. $\frac{1}{2}$ $\frac{2}{5}$ $\frac{1}{4}$ ____________
14. $\frac{3}{18}$ $\frac{5}{6}$ $\frac{2}{3}$ ____________
15. $\frac{2}{5}$ $\frac{6}{7}$ $\frac{5}{9}$ ____________

Name ____________

LOST!

Badminton may seem like a rather easy sport where you just hit the "birdie" around at a slow pace. Actually, it is the world's fastest racket sport. The "birdies" are really called shuttlecocks, and they travel as fast as 200 miles per hour. Players must be very quick, strong, and agile to compete.

Pete has gotten separated from the badminton team on the way to the competition. To help him join his teammates, compare the fractions in each box. Color the boxes that have the correct sign (<, >, or =) between the fractions. If you do this correctly, you will have colored a path for Pete.

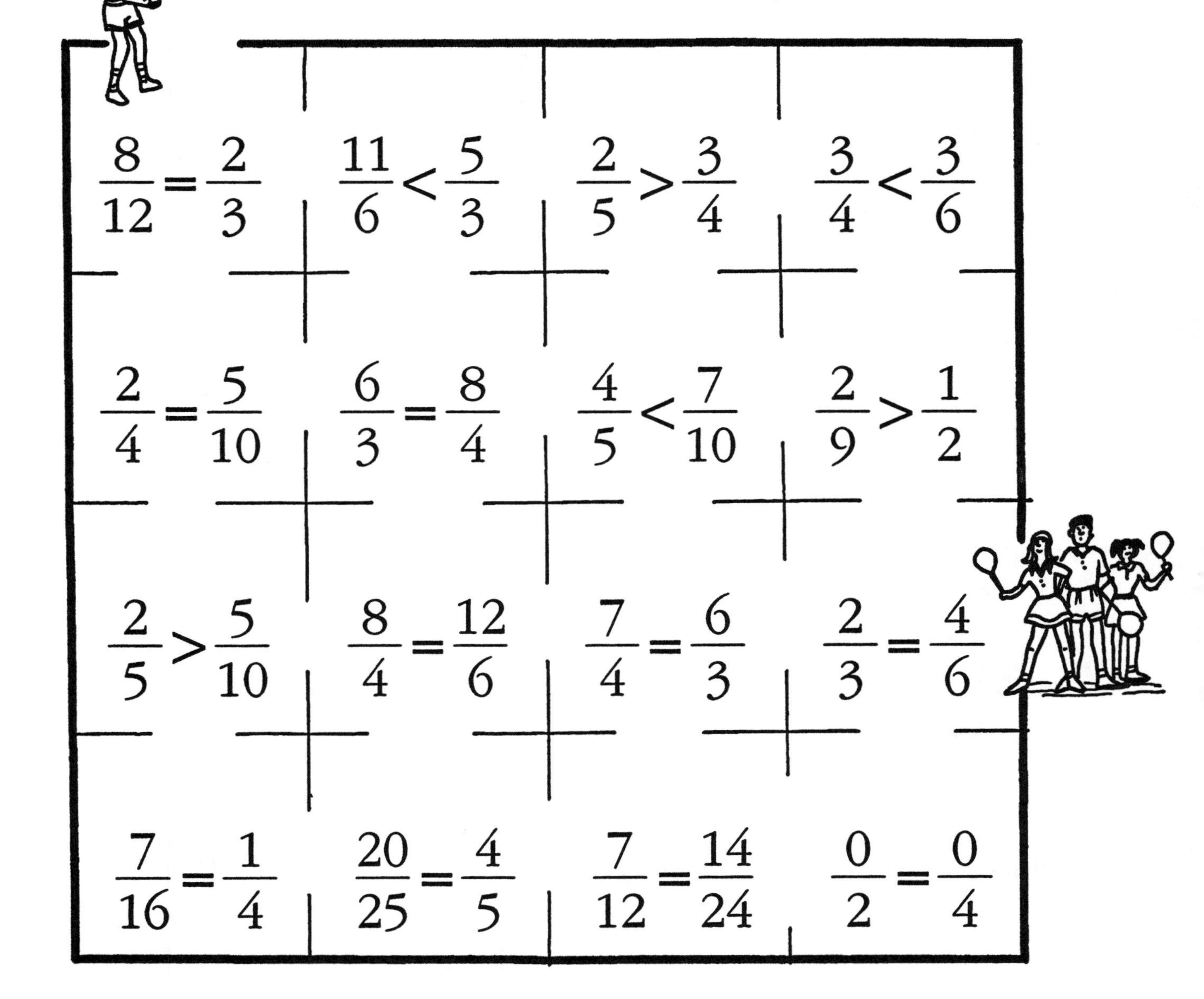

Name

Equivalent Fractions

WINTER OLYMPIC TRIVIA

Do you know the name of the most difficult ice-skating jump ever landed in Olympic competition? Do you know what is the oldest game played on ice? Do you know how fast downhill skiers might travel? Do you know how many people fit on a luge sled? Do you know the length of the longest cross-country ski race?

Find the answers to these and other trivia questions while you practice identifying equivalent fractions. In each problem, two of the fractions are equivalent. The fraction that is not equivalent gives the answer to the trivia question! Circle the non-equivalent fraction in each problem.

1. Luge sleds can reach speeds over
   A. $\frac{16}{18}$ 150 mph
   B. $\frac{8}{9}$ 300 mph
   C. $\frac{5}{9}$ 80 mph

2. The oldest game played on ice is
   A. $\frac{2}{3}$ curling
   B. $\frac{1}{5}$ ice hockey
   C. $\frac{6}{9}$ ice bowling

3. Downhill racers travel at speeds of up to
   A. $\frac{4}{5}$ 200 mph
   B. $\frac{7}{9}$ 80 mph
   C. $\frac{12}{15}$ 40 mph

4. The number of competitors riding each luge sled is
   A. $\frac{3}{5}$ 1 or 2
   B. $\frac{4}{7}$ 3 or 4
   C. $\frac{8}{14}$ 4 or 5

5. The first Olympics that included snowboarding was in
   A. $\frac{1}{3}$ 1992
   B. $\frac{7}{21}$ 1984
   C. $\frac{5}{8}$ 1998

6. The speedskating rink in Lillehammer in 1994 was shaped like a
   A. $\frac{7}{8}$ ice skate
   B. $\frac{9}{12}$ Viking ship
   C. $\frac{28}{32}$ snowshoe

7. People have been using skis for
   A. $\frac{1}{11}$ 9000 years
   B. $\frac{2}{12}$ 200 years
   C. $\frac{1}{6}$ 100 years

8. How far can ski jumpers fly?
   A. $\frac{3}{4}$ about 600 feet
   B. $\frac{6}{7}$ about 1 mile
   C. $\frac{18}{21}$ about 2000 feet

9. The biathlon combines
   A. $\frac{1}{2}$ skating & skiing
   B. $\frac{5}{11}$ cross-country skiing & rifle shooting
   C. $\frac{2}{4}$ luge & bobsled

10. The most difficult ice-skating jump landed in Olympic competition (as of 1997) was
   A. $\frac{1}{4}$ the quadruple lutz
   B. $\frac{2}{8}$ the triple flip
   C. $\frac{2}{6}$ the triple axle

Name

THE LONGEST JUMPS

It sounds pretty hard! An athlete runs down a short path and jumps as far as possible, landing into a pit of sand. A measurement is taken from the beginning of the jump to the impression the body leaves in the sand. If the athlete falls backward from where the feet land, the measurement will be shorter than desired!

Here are some measurements of long jumps from athletes of all ages. They are written as improper fractions. Change them into mixed numerals.

1. Carl $\frac{57}{2}$ feet = ______________
2. Lutz $\frac{57}{6}$ feet = ______________
3. Jackie $\frac{97}{4}$ feet = ______________
4. Heike $\frac{47}{2}$ feet = ______________
5. Amber $\frac{32}{5}$ feet = ______________
6. Yvette $\frac{85}{8}$ feet = ______________
7. Arnie $\frac{88}{3}$ feet = ______________
8. Ellery $\frac{83}{4}$ feet = ______________
9. James $\frac{49}{4}$ feet = ______________
10. Randy $\frac{109}{4}$ feet = ______________
11. Tatyana $\frac{71}{3}$ feet = ______________
12. Mary $\frac{63}{4}$ feet = ______________
13. Bob $\frac{165}{6}$ feet = ______________
14. Albert $\frac{129}{12}$ feet = ______________
15. Jenny $\frac{101}{4}$ feet = ______________
16. Tommy $\frac{14}{3}$ feet = ______________

Name ______________________________

GETTING TO VENUES

A venue is a place where one of the Olympic events is held. There are many venues at each Olympic Games. These Olympic athletes are trying to get to their proper venues, but their paths are blocked. Remove the obstacles along the paths by changing each improper fraction to its correct mixed numeral.

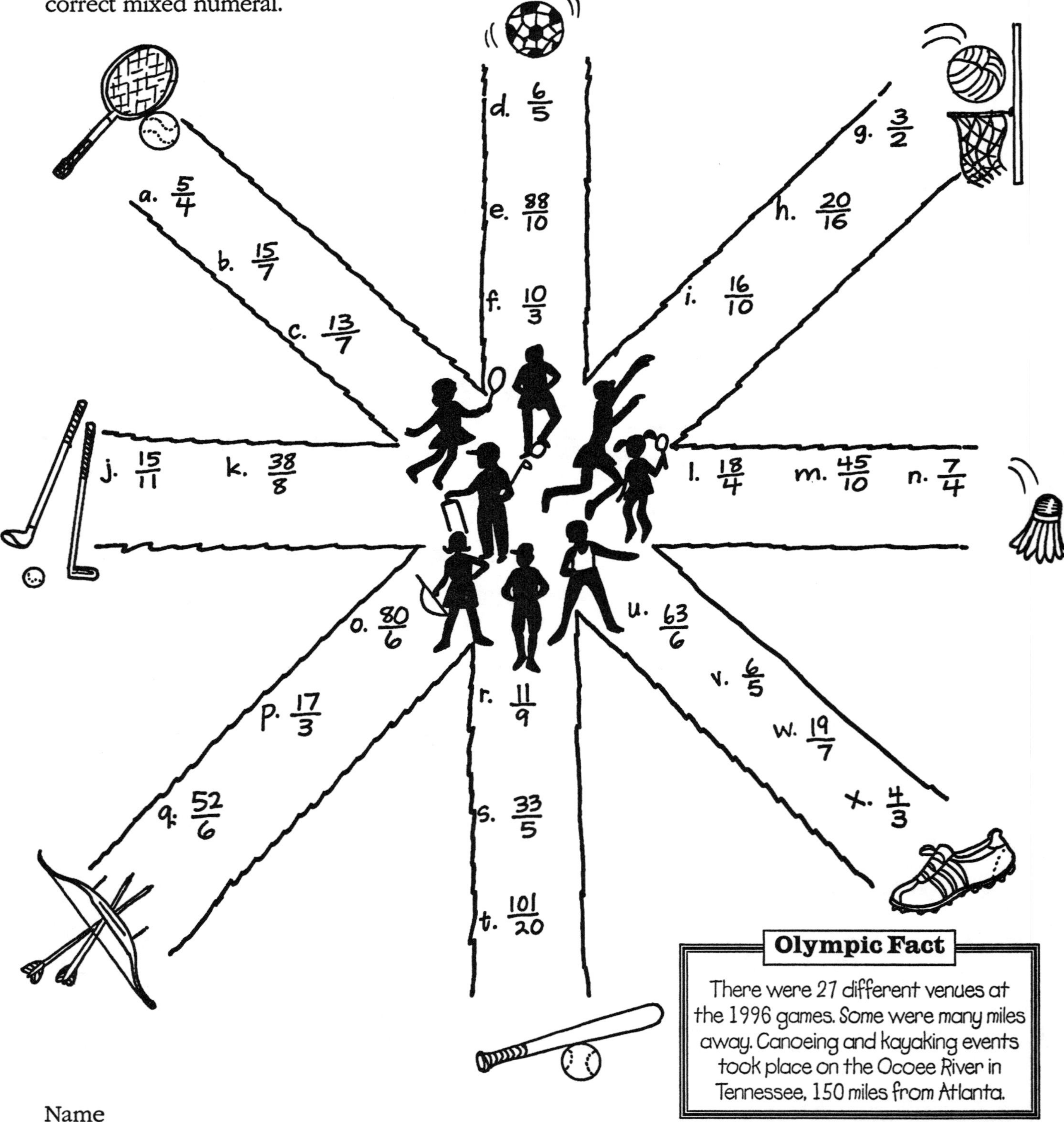

Olympic Fact

There were 27 different venues at the 1996 games. Some were many miles away. Canoeing and kayaking events took place on the Ocoee River in Tennessee, 150 miles from Atlanta.

Name

THE #1 SPORT

In ancient versions of soccer, players tossed the ball around in the air, bouncing it off their hands and heads. Today, only the goalie is allowed to touch the ball with his or her hands while it is in play on the field.

Soccer was the first team sport to be included in the Olympics. At every Olympic Games, it draws some of the biggest crowds. In Barcelona, Spain, the mainly Spanish crowd was thrilled to see the Spanish team win the gold medal!

Look on the soccer field for the answer to each problem. Circle the correct answer with the color shown next to the problem. Answers must be in lowest terms.

Example: $\frac{1}{10} + \frac{1}{2} = \frac{1}{10} + \frac{5}{10} = \frac{6}{10}$ ($\frac{3}{5}$ *in lowest terms)*

1. GREEN: $\frac{2}{3} + \frac{1}{6} =$ ________
2. RED: $\frac{5}{10} - \frac{1}{5} =$ ________
3. BLUE: $\frac{5}{12} - \frac{1}{3} =$ ________
4. YELLOW: $\frac{3}{4} - \frac{5}{8} =$ ________
5. PURPLE: $\frac{1}{4} + \frac{4}{16} =$ ________
6. BROWN: $\frac{10}{25} + \frac{2}{5} =$ ________
7. ORANGE: $\frac{11}{12} - \frac{3}{4} =$ ________
8. PINK: $\frac{1}{2} + \frac{2}{22} =$ ________
9. RED: $\frac{20}{30} - \frac{2}{6} =$ ________
10. BLUE: $\frac{1}{9} + \frac{2}{3} - \frac{1}{3} =$ ________
11. PURPLE: $\frac{2}{9} + \frac{8}{9} - \frac{1}{3} =$ ________
12. GREEN: $\frac{4}{7} + \frac{1}{3} =$ ________
13. ORANGE: $\frac{11}{14} - \frac{3}{7} + \frac{1}{7} =$ ________
14. BROWN: $\frac{1}{6} + \frac{3}{4} - \frac{1}{8} =$ ________

$\frac{19}{21}$ $\frac{13}{22}$ $\frac{7}{9}$ $\frac{4}{9}$

$\frac{1}{8}$ $\frac{5}{6}$ $\frac{1}{3}$ $\frac{3}{10}$

$\frac{1}{2}$ $\frac{1}{12}$ $\frac{4}{5}$ $\frac{1}{6}$ $\frac{1}{2}$ $\frac{19}{24}$

Name

PENTATHLON CALCULATIONS

Penta means five, so athletes who compete in the pentathlon have to be good at five different sports. The modern pentathlon is based on the duties of a warrior who must deliver a message across enemy lines. He has to ride a horse around many obstacles, defend himself with a sword and gun, run great distances, and swim across rivers and streams. Olympic competitors must complete contests in equestrian riding, fencing, pistol shooting, running, and swimming.

Each of these calculations has five parts, also. You need to be good at each step in order to get the right answer!

1. $\frac{9}{10} - \frac{1}{10} + \frac{7}{10} + \frac{5}{10} - \frac{1}{10} =$ ______________

2. $\frac{7}{9} - \frac{3}{9} - \frac{2}{9} + \frac{6}{9} - \frac{2}{9} =$ ______________

3. $\frac{2}{13} + \frac{5}{13} - \frac{4}{13} + \frac{10}{13} - \frac{6}{13} =$ ______________

4. $\frac{5}{6} - \frac{2}{6} + \frac{6}{6} - \frac{2}{6} - \frac{3}{6} =$ ______________

5. $\frac{3}{20} - \frac{1}{20} + \frac{15}{20} - \frac{4}{20} + \frac{2}{20} =$ ______________

6. $\frac{1}{11} + \frac{5}{11} + \frac{8}{11} - \frac{9}{11} + \frac{1}{11} =$ ______________

7. $\frac{1}{5} - \frac{1}{5} + \frac{2}{5} + \frac{9}{5} - \frac{6}{5} =$ ______________

8. $\frac{4}{16} - \frac{2}{16} + \frac{7}{16} + \frac{5}{16} + \frac{1}{16} =$ ______________

9. $\frac{5}{25} - \frac{3}{25} + \frac{9}{25} - \frac{2}{25} + \frac{1}{25} =$ ______________

10. $\frac{6}{12} - \frac{5}{12} + \frac{7}{12} - \frac{7}{12} + \frac{1}{12} =$ ______________

11. $\frac{9}{6} - \frac{2}{6} + \frac{10}{6} - \frac{3}{6} + \frac{15}{6} =$ ______________

12. $\frac{11}{100} + \frac{15}{100} - \frac{4}{100} + \frac{50}{100} - \frac{1}{100} =$ ______________

13. $\frac{3}{30} + \frac{8}{30} + \frac{14}{30} - \frac{7}{30} - \frac{2}{30} =$ ______________

14. $999\frac{8}{10} - 999 + \frac{14}{10} - \frac{2}{10} + \frac{5}{10} =$ ______________

Name

WHEN IS A SHELL A SCULL?

An exciting sport in the Summer Olympic Games is rowing. Boats used for competitive rowing are called shells. Some of the shells are sculls, but some are not! It depends on how the oars are arranged. In sculls, each crew member rows with two oars instead of one—that's the main difference! The boats are very light and move quickly over the 2000-meter course.

Solve each row of problems as quickly as the crews row the shells. Don't be sloppy, or it will take longer to correct your mistakes! Can you tell which boat is the scull?

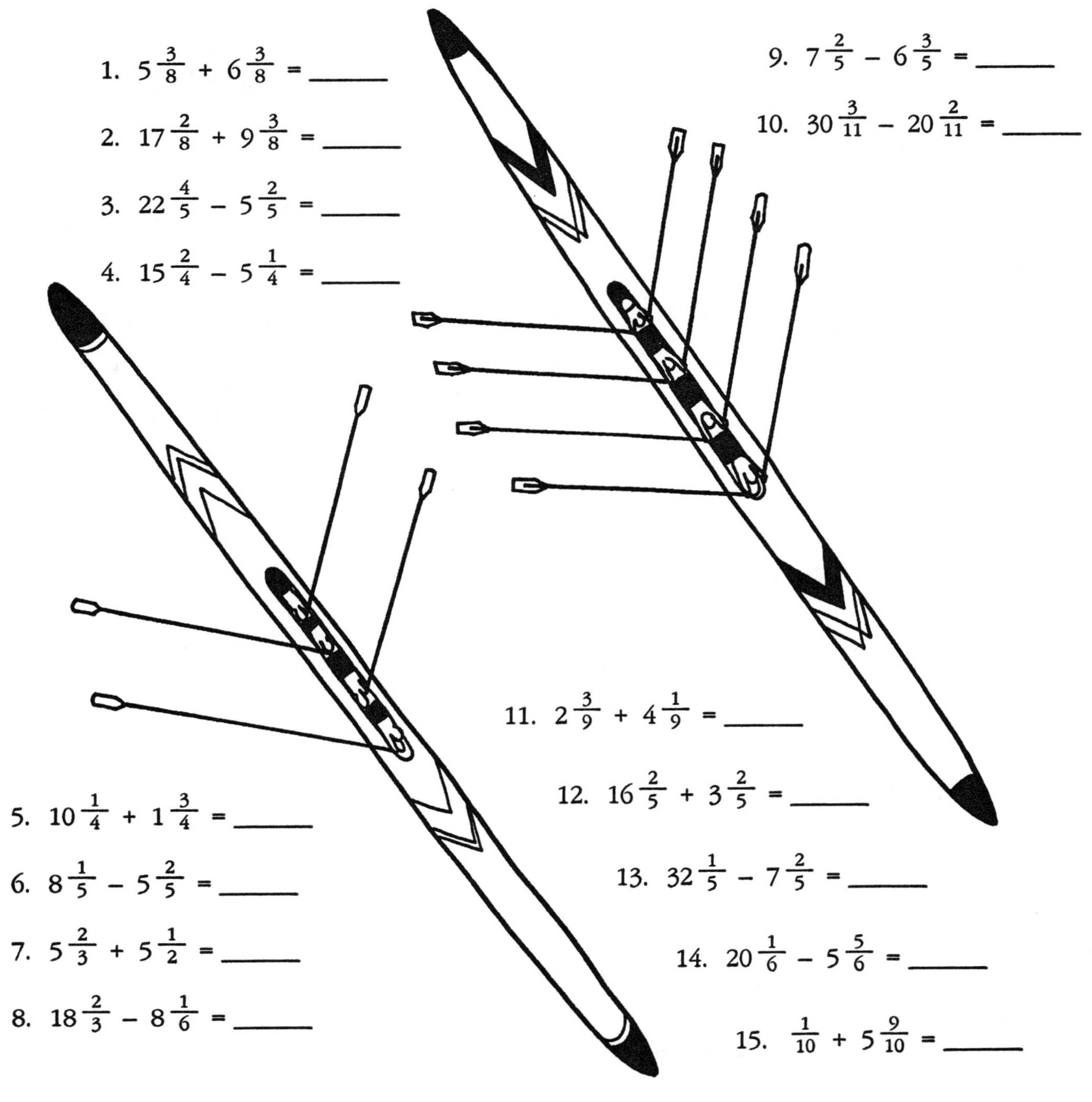

1. $5\frac{3}{8} + 6\frac{3}{8} =$ ______
2. $17\frac{2}{8} + 9\frac{3}{8} =$ ______
3. $22\frac{4}{5} - 5\frac{2}{5} =$ ______
4. $15\frac{2}{4} - 5\frac{1}{4} =$ ______
5. $10\frac{1}{4} + 1\frac{3}{4} =$ ______
6. $8\frac{1}{5} - 5\frac{2}{5} =$ ______
7. $5\frac{2}{3} + 5\frac{1}{2} =$ ______
8. $18\frac{2}{3} - 8\frac{1}{6} =$ ______
9. $7\frac{2}{5} - 6\frac{3}{5} =$ ______
10. $30\frac{3}{11} - 20\frac{2}{11} =$ ______
11. $2\frac{3}{9} + 4\frac{1}{9} =$ ______
12. $16\frac{2}{5} + 3\frac{2}{5} =$ ______
13. $32\frac{1}{5} - 7\frac{2}{5} =$ ______
14. $20\frac{1}{6} - 5\frac{5}{6} =$ ______
15. $\frac{1}{10} + 5\frac{9}{10} =$ ______

Name ______

HOT OLYMPIC STEW!

Twenty-four cold, hungry skiers stopped at the ski lodge to warm up with some hearty Olympic Stew. The cook's recipe was intended to serve 36. She knew that she would need to make only $\frac{2}{3}$ of her recipe.

Rewrite the stew recipe to help the cook make the smaller batch. Multiply each amount from the original recipe by $\frac{2}{3}$ to find the new amount.

OLYMPIC STEW FOR 36	OLYMPIC STEW FOR 24
$5\frac{1}{2}$ pounds potatoes	
$8\frac{1}{4}$ quarts boiling water	
2 large onions	
$8\frac{1}{8}$ cups chicken broth	
$4\frac{2}{3}$ carrots, chopped	
$8\frac{1}{2}$ celery sticks, sliced	
$1\frac{1}{2}$ green peppers, chopped	
$5\frac{1}{3}$ cups frozen corn	
$4\frac{3}{4}$ pounds mushrooms	
$7\frac{1}{4}$ cups cooked chicken, diced	
$3\frac{1}{3}$ teaspoons salt	
$6\frac{1}{3}$ Tablespoons mixed herbs	
Mix all ingredients in a large pot. Cook over medium heat for one hour, stirring often.	

Name

THROUGH WILD WATERS

In the Olympic kayaking events, kayakers race through wild, foaming water (called white-water). They must get down the river through a series of gates safely and fast! Some of the gates require them to paddle upstream against the raging waters! Of course, sometimes the kayaks flip, but the athletes are good at turning right side up again.

To divide fractions, you need to do some flipping, too! The second number in the problem must be turned upside down. Then, you multiply the two fractions to get the answer to the division problem!

$$\frac{3}{5} \div \frac{7}{10} = \frac{3}{5} \times \frac{10}{7} = \frac{30}{35} = \frac{6}{7}$$

Flip the second fraction in all these problems to find the right answers.

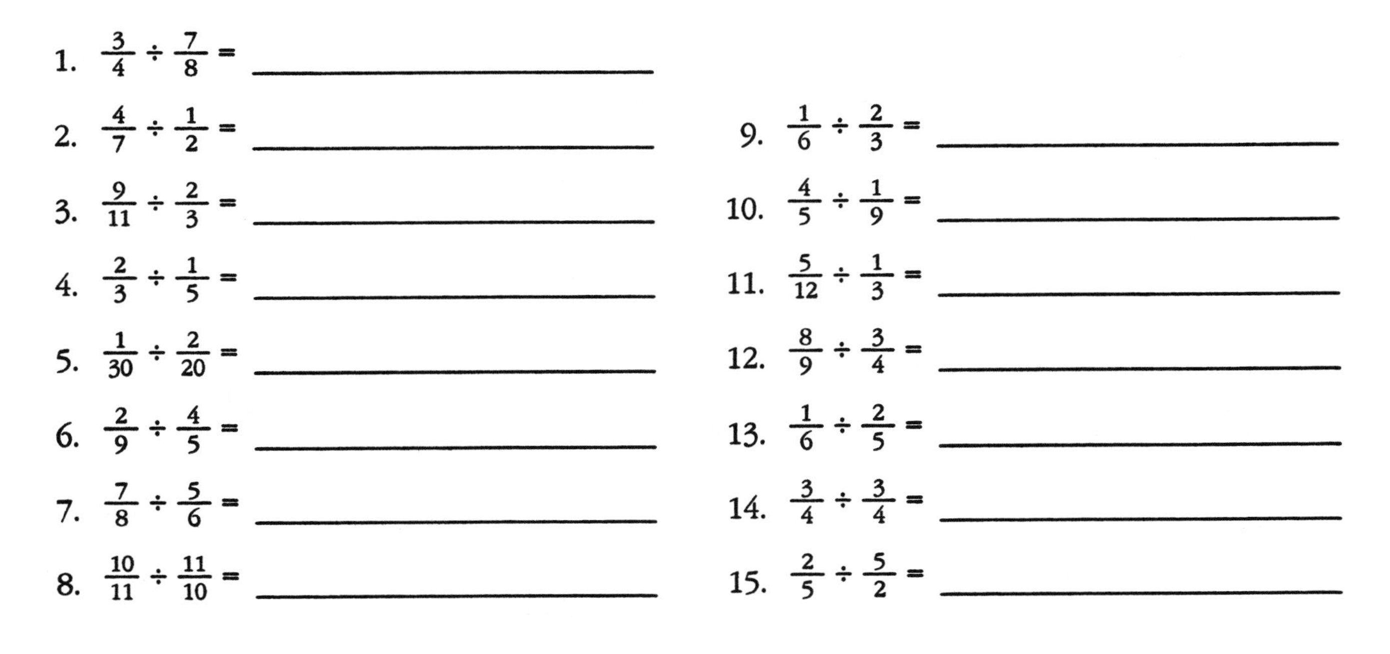

1. $\frac{3}{4} \div \frac{7}{8} =$ ____________
2. $\frac{4}{7} \div \frac{1}{2} =$ ____________
3. $\frac{9}{11} \div \frac{2}{3} =$ ____________
4. $\frac{2}{3} \div \frac{1}{5} =$ ____________
5. $\frac{1}{30} \div \frac{2}{20} =$ ____________
6. $\frac{2}{9} \div \frac{4}{5} =$ ____________
7. $\frac{7}{8} \div \frac{5}{6} =$ ____________
8. $\frac{10}{11} \div \frac{11}{10} =$ ____________
9. $\frac{1}{6} \div \frac{2}{3} =$ ____________
10. $\frac{4}{5} \div \frac{1}{9} =$ ____________
11. $\frac{5}{12} \div \frac{1}{3} =$ ____________
12. $\frac{8}{9} \div \frac{3}{4} =$ ____________
13. $\frac{1}{6} \div \frac{2}{5} =$ ____________
14. $\frac{3}{4} \div \frac{3}{4} =$ ____________
15. $\frac{2}{5} \div \frac{5}{2} =$ ____________

Name ____________

Read Decimals

TAKE THE PLUNGE!

Can you imagine jumping off a three-story building into a pool of water? This is what platform divers do. Olympic divers either jump off high platforms, where they begin at a standstill, or they jump off a bouncy springboard. Seven judges watch each dive and score it between 0 and 10. Scores for eleven dives are added together. The diver with the highest score wins. In 1982, Mary Ellen Clark got a bronze medal for the USA with a score of 401.91.

Find a decimal in the pool to match each of the decimal words below.

1. one hundred eighty-three ten thousandths ____________
2. one and eighty-three thousandths ____________
3. one hundred fifteen thousandths ____________
4. fifty-five hundredths ____________
5. ninety-nine and seven tenths ____________
6. thirteen and four tenths ____________
7. five hundred and five thousandths ____________
8. nine and seventy-eight hundredths ____________
9. ninety-seven hundredths ____________
10. ten and eighty-three hundredths ____________
11. one hundred eight and three tenths ____________
12. two hundred thirty-four thousandths ____________
13. five and five hundred fifty-five thousandths ____________
14. nine hundred seventy-eight thousandths ____________
15. nine hundred seventy-eight and three tenths ____________
16. eleven and five hundred one thousandths ____________

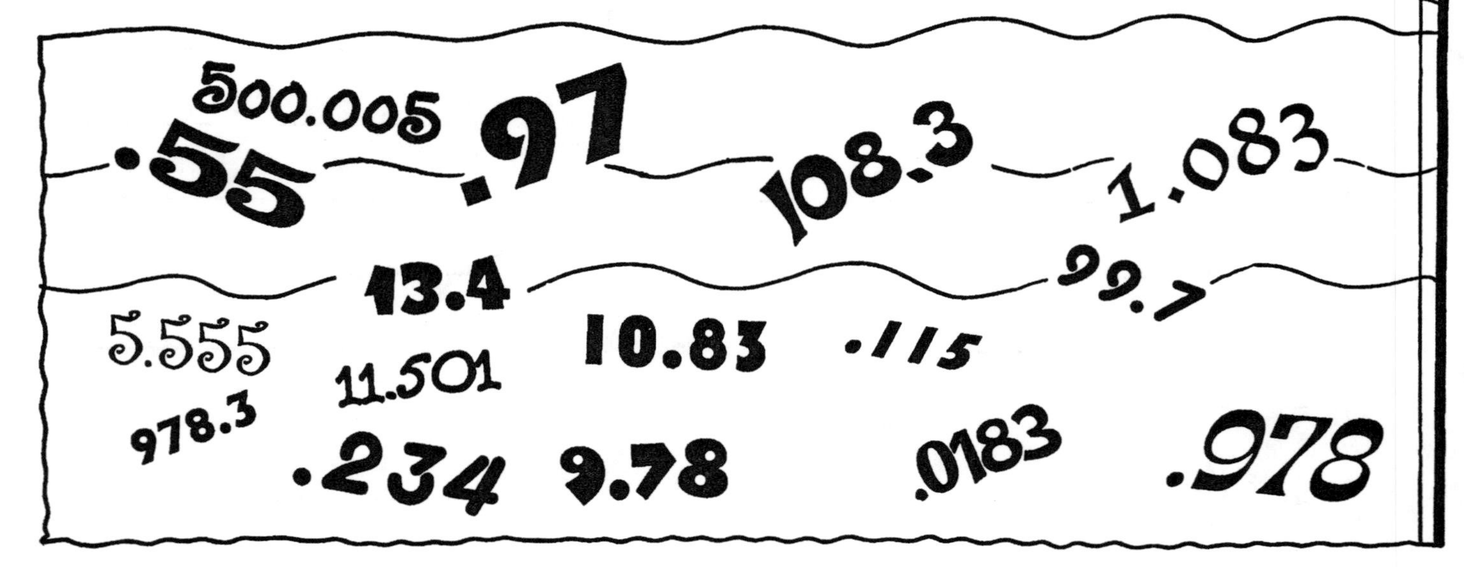

Name

NOSE ROLLS & FAKIES

This must be the sport with the wildest names for moves and tricks! On a snowboard you can do Halfpipes, Nose Rolls, Wheelies, McTwists, Chicken Salads, and Ollies—and many more tricks with wild, wacky names! 1998 was the first time snowboarders could take part in the Olympic Games. The boarders were ready to do all these fancy tricks, and more, in Japan!

To finish each of these tricks with a good score, read the decimals on each card. Then number them in order from the largest to the smallest.

Trick #1 FAKIE

____	0.11103	____	1.7
____	0.103	____	11.3
____	10.37	____	13.01
____	11.370	____	0.13

Trick #2 NOSE ROLL

____	15.02	____	15.21
____	1.5	____	1.51
____	0.005	____	55.5
____	0.05	____	5.5

Trick #3 BACKSCRATCHER

____	4.5	____	4.7
____	0.451	____	44.5
____	0.45	____	4.4
____	0.06	____	0.44

Trick #4 McTWIST

____	5.28	____	5.6
____	9.97	____	0.009
____	0.8	____	5.8
____	0.99	____	0.08

Trick #5 CHICKEN SALAD

____	2.6	____	6.2
____	2.7	____	2.9
____	2.06	____	22.6
____	26.6	____	2.999

Trick #6 OLLIE

____	7.2	____	7.7
____	0.72	____	77.27
____	0.072	____	0.07
____	72.1	____	0.007

Trick #7 TAIL WHEELIE

____	0.0001	____	0.000001
____	0.001	____	0.01
____	101.1	____	10.11
____	1.1	____	0.00011

Name

A HUGE OBSTACLE COURSE

So many obstacles! A runner in the steeplechase race has to run 3,000 meters and jump over 28 hurdles and 7 water jumps. See if you can get past all the obstacles in this steeplechase course. At each jump, round the decimal as the directions tell you. If you get them all correct, you will have successfully completed this steeplechase course. The real Olympic course will be a lot harder than this!

Round to the nearest tenth.

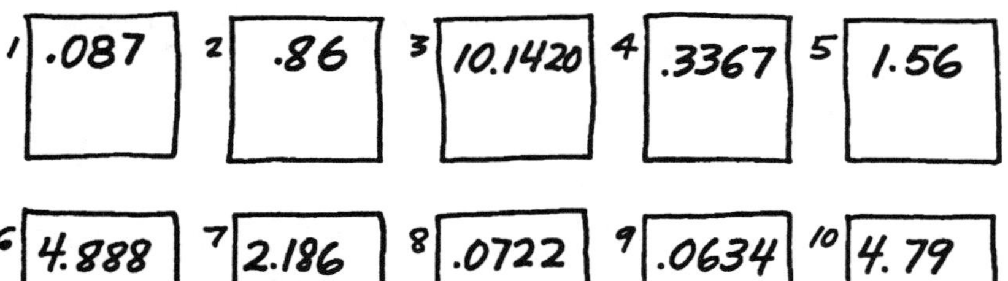

Round to the nearest hundredth.

Round to the nearest thousandth.

21 .4689
22 4.6789
23 7.08966
24 .05555
25 41.5226
26 .198765
27 5.0109
28 .02161

Round to the nearest ten thousandth.

29 .747777
30 .19991
31 .740000
32 15.02891
33 1.1515151
34 4.33371
35 .59022

Olympic Fact

Larissa Latynina, a gymnast from the USSR, holds the record for the most medals won ever—18. She also won 9 gold medals—the most ever for a woman.

Name

WHO WEARS THE MEDALS?

In the Olympics, the individual all-around championship is the highest achievement a gymnast can achieve. Most gymnasts dream of winning this gold medal. Gymnasts must compete in four events. Their scores from all four events are totaled to see who has the highest score.

Add up the scores for all these gymnasts. Then rank them in order from first to last.

Gymnast	Balance Beam	Floor Exercise	Uneven Bars	Horse Vault	Total Score	Place
Karin	9.932	9.912	9.955	9.680		
Sofia	9.817	9.950	9.609	9.896		
Elena	8.954	9.987	9.640	9.320		
Kim	8.999	9.690	9.800	9.975		
Kerri	9.981	9.208	9.997	9.700		
Tatiana	9.975	10.00	9.980	9.973		
Nina	9.290	9.964	9.699	9.609		
Larissa	9.956	9.866	9.057	9.666		
Svetlana	9.979	9.979	9.780	10.00		
Olga	8.974	9.401	8.899	9.789		
Kathy	10.00	9.777	9.780	9.925		
Tamara	9.966	10.00	9.224	9.099		

Who won the Gold? ______________ Silver? ______________ Bronze? ______________

Name ____

WHAT'S THE COST?

Sonja Henie was eleven years old when she entered her first Olympic Games in 1924. Even though this young figure skater finished last, she did not give up. She came back three more times and won the gold medal every time! She was known for her interesting, graceful movements and her fancy costumes. Those fancy costumes and other supplies add up to a lot of expense for a skater! You can be sure that they are all more expensive today than they were in Sonja Heine's time! Practice your decimal skills to find the costs for these skating items. Use scrap paper to solve the problems.

_______ 1. One skater paid $108.00 for 36 fancy jewels to sew on her costume. What did each jewel cost?

_______ 2. Laces for her skates were 5 pair for $13.00. What does one pair cost?

_______ 3. A pair of skate blades costs $189.00. A pair of skate boots costs 4 times that much. How much are the skates and blades all together?

_______ 4. If a skater's vitamins for one month cost $17.50, how much does one year's supply cost?

_______ 5. Every practice session at Kurt's rink costs $5.00. Kurt goes to 4 sessions a day, 6 days a week. How much does he spend each week on ice time?

_______ 6. If Jill's skating tights cost $53.60 for 8 pair, how much will 4 pair cost?

_______ 7. The coach's fees are $45.00 per hour. Jill trains with her coach 8 hours each week. How much per week does this cost Jill?

_______ 8. Last year, Scott paid $800 in entry fees for 6 competitions. If each fee was the same, about how much did each competition cost him to enter?

_______ 9. If Kristi's new skates cost $695 and she buys 3 pair each year, how much would she spend in a year on skates?

_______ 10. If Paul spent $322.00 on moleskin and cream for blisters last year, how much did it cost him per month to take care of blisters?

_______ 11. Jenni's newest costume cost twice as much as her last one. This one was $286. How much did the last one cost?

_______ 12. Todd's skating partner drinks hot chocolate twice a day at the rink. The hot chocolate costs $1.25, and they skated 290 days last year. How much did she spend on hot chocolate all year?

Name ____

OVER THE TOP

Pole vaulters sprint along a short track with a long, flexible pole. Then they plant the pole and soar upside down over another pole that might be almost 20 feet high. The goal is to make it over the top without knocking off that pole! At the 1996 Olympics, Jean Galfione from France won the gold medal with a jump over a pole that was 19 feet, 5 inches high!

If a pole vaulter makes it over the top 6 times out of 7 tries, a fraction ($\frac{6}{7}$) can show his success rate. The fraction can be changed to a decimal score. (Divide 6 by 7. The decimal is 0.86.) Find the decimal to match each fraction that shows how these pole vaulters are doing at their practice. Round to the nearest hundredth.

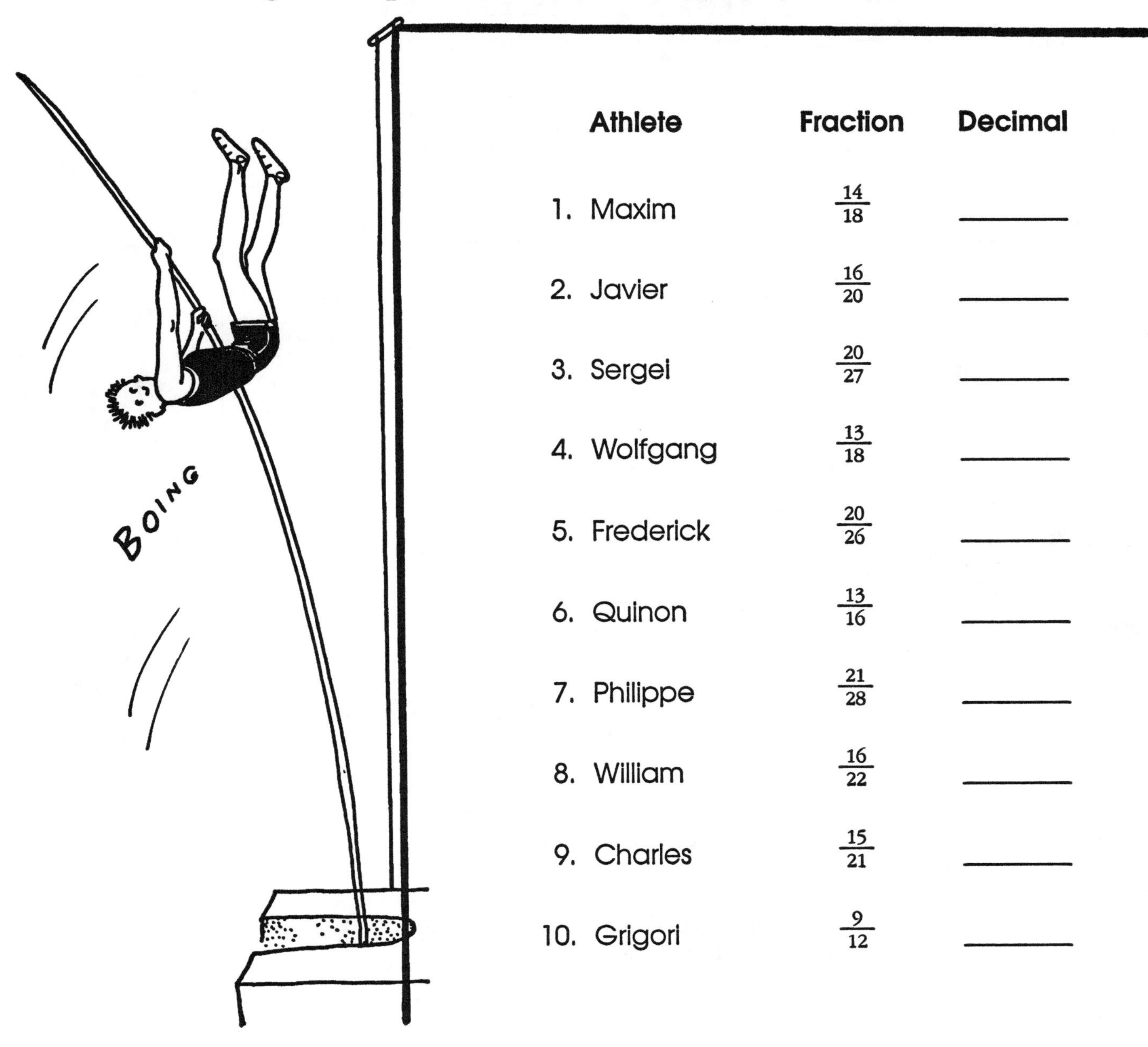

Athlete	Fraction	Decimal
1. Maxim	$\frac{14}{18}$	________
2. Javier	$\frac{16}{20}$	________
3. Sergei	$\frac{20}{27}$	________
4. Wolfgang	$\frac{13}{18}$	________
5. Frederick	$\frac{20}{26}$	________
6. Quinon	$\frac{13}{16}$	________
7. Philippe	$\frac{21}{28}$	________
8. William	$\frac{16}{22}$	________
9. Charles	$\frac{15}{21}$	________
10. Grigori	$\frac{9}{12}$	________

Name ______________________________

THE DREAM TEAM

No one doubted that the United States Olympic basketball team would win gold in Barcelona in 1992. This was a team of the world's best professional players, including Magic Johnson and Michael Jordan. These players probably had pretty high percentages when it came to shooting free throws.

Each fraction shows the number of free throws that might have been made by some basketball players from the top three Olympic teams in comparison to the number of shots taken. Change each fraction into a percentage.

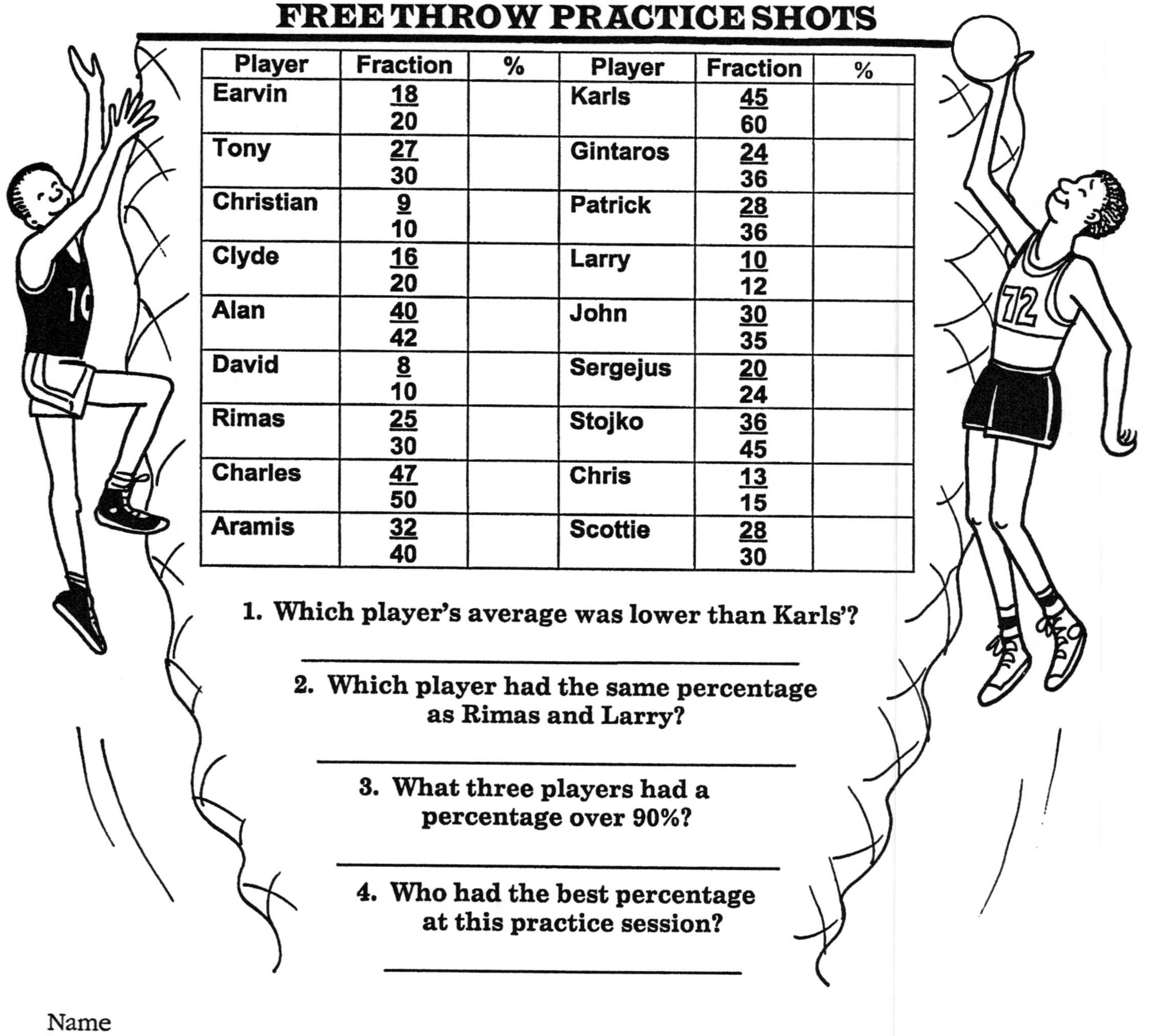

FREE THROW PRACTICE SHOTS

Player	Fraction	%	Player	Fraction	%
Earvin	$\frac{18}{20}$		Karls	$\frac{45}{60}$	
Tony	$\frac{27}{30}$		Gintaros	$\frac{24}{36}$	
Christian	$\frac{9}{10}$		Patrick	$\frac{28}{36}$	
Clyde	$\frac{16}{20}$		Larry	$\frac{10}{12}$	
Alan	$\frac{40}{42}$		John	$\frac{30}{35}$	
David	$\frac{8}{10}$		Sergejus	$\frac{20}{24}$	
Rimas	$\frac{25}{30}$		Stojko	$\frac{36}{45}$	
Charles	$\frac{47}{50}$		Chris	$\frac{13}{15}$	
Aramis	$\frac{32}{40}$		Scottie	$\frac{28}{30}$	

1. Which player's average was lower than Karls'?

____

2. Which player had the same percentage as Rimas and Larry?

____

3. What three players had a percentage over 90%?

____

4. Who had the best percentage at this practice session?

____

Name

A GREAT MATCH

Tennis originated in the thirteenth century in France. Players would stretch a cord of rope across a room and hit a cloth bag full of hair back and forth over the rope! Now tennis is an Olympic sport with four events that include singles and doubles for men and women.

Olympic Fact

Tennis balls are made from rubber molded into 2 cups that are cemented together and covered with wool. A new tennis ball can bounce about 55 feet.

For each percentage shown on one side of this tennis court, find a matching fraction from the other side. Use a different colored marker to draw lines connecting the pairs.

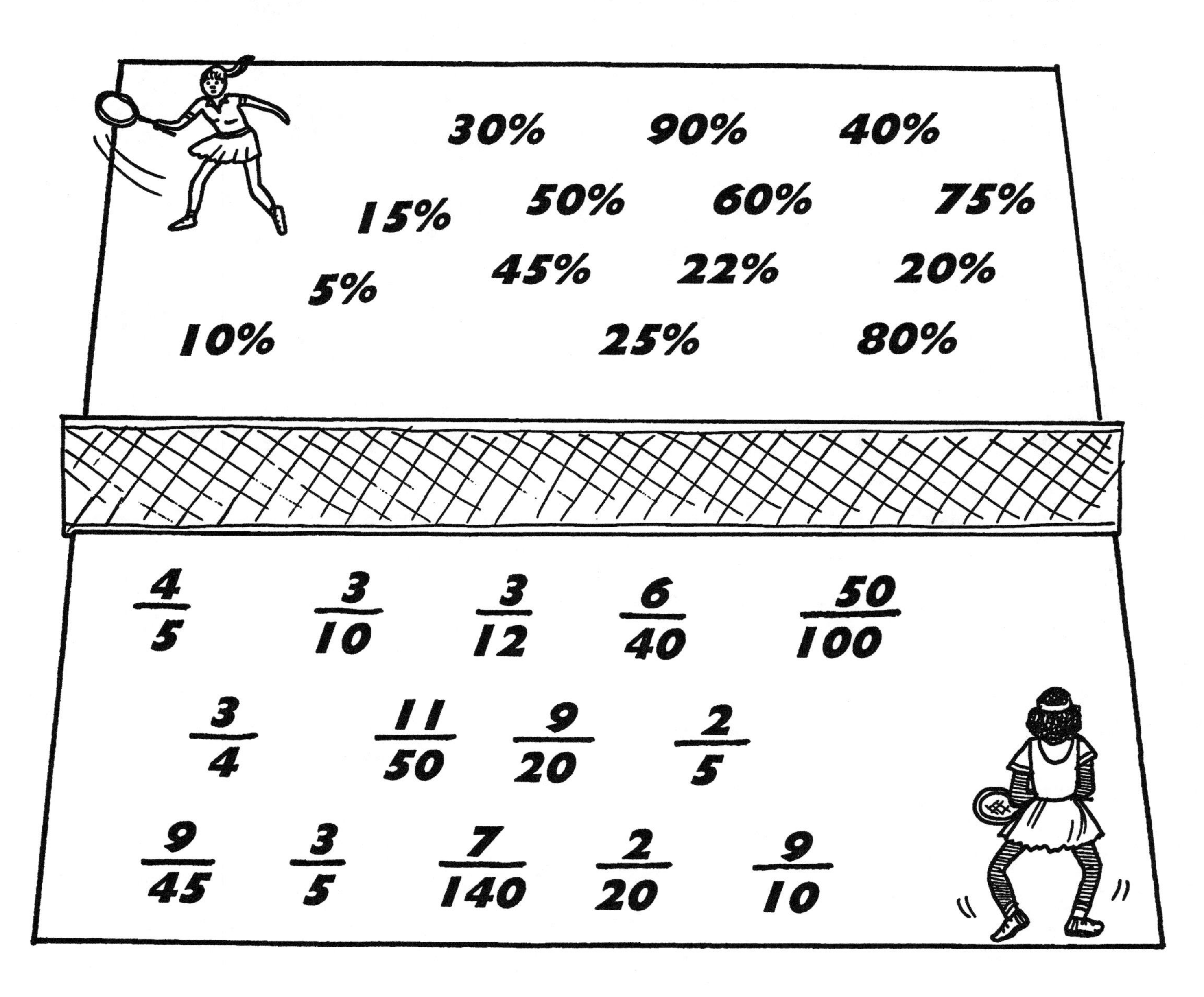

Name

THE FINAL COUNT

When the Olympic Games are over, the medals are counted. At the 1996 Summer Olympic Games in Atlanta, 842 medals were awarded. This is the way the final count looked for the top 20 medal-winning countries. Use the chart to solve the problems below.

Summer Olympic Games, 1996
Final Medal Standings for Top 20 Countries

Country	Gold	Silver	Bronze	Total Medals
United States	44	32	25	
Germany	20	18		65
Russia		21	16	63
China	16	22	12	
Australia	9		23	41
France		7	15	37
Italy	13	10	12	
South Korea	7	15		27
Cuba	9	8		25
Ukraine		2	12	23
Canada	3		8	22
Hungary		4	10	21
Romania	4		9	20
Netherlands	4	5		19
Poland	7	5	5	
Spain	5	6	6	
Bulgaria	3	7	5	
Brazil	3		9	15
Great Britain	1	8	6	
Belarus	1	6		15

1. Write the missing numbers in the spaces on the chart.

2. Which four countries won the same total number of medals? ____________________

____

3. Which two countries won 17 medals?

____

4. Which three countries each won 12 bronze medals? ____________________

____

____

5. Which country won the same number of gold and bronze medals? ____________

6. Which country won 8 times as many bronze medals as gold medals? ________

7. Which country won twice as many medals as Cuba? ____________________

8. Which country won more bronze medals than the U.S.? ________________

9. How many more gold medals did Russia win than Ukraine? ______________

10. Which country won half of its medals in silver? ____________________

Use with page 51.

Name

THE FINAL COUNT, CONT.

The medal count is quite different at the Winter Olympics because there are fewer events. The chart below tells the final medal count for all medal-winning countries at the 1994 Winter Olympic Games in Lillehammer, Norway. Use the information to solve the problems below.

1. Write the missing numbers in the spaces on the chart.

_____________ 2. Find the total number of medals awarded in 1994.

_____________ 3. How many medals did the top 6 countries win?

_____________ 4. How many medals did the other 16 countries win?

_____________ 5. Which country won more gold medals than Norway?

_____________ 6. Which country won more silver and bronze, but fewer gold medals than the U.S.?

_____________ 7. Which country won three times the gold medals of Canada?

_____________ 8. Which country won the same number of gold medals as Switzerland?

_____________ 9. Which country won 17 more medals than China?

_____________ 10. Which country won 22 fewer medals than Norway?

Winter Olympic Games, 1994
Final Medal Standings

Country	Gold	Silver	Bronze	Total Medals
Norway	10	11	5	
Germany	9	7		24
Russia	11		4	23
Italy	7	5	8	
United States	6		2	13
Canada		6	4	13
Switzerland		4	2	9
Austria	2	3	4	
South Korea		1	1	6
Finland	0		5	6
Japan	1	2		5
France	0		4	5
Netherlands	0	1	3	
Sweden		1	0	3
Kazakstan	1	2	0	
China	0		2	3
Slovenia	0	0		3
Ukraine		0	1	2
Belarus	0	2	0	
Great Britain	0		2	2
Uzbekistan	1	0		1
Australia		0	1	1

Use with page 50.

Name

TEMPERATURE COUNTS

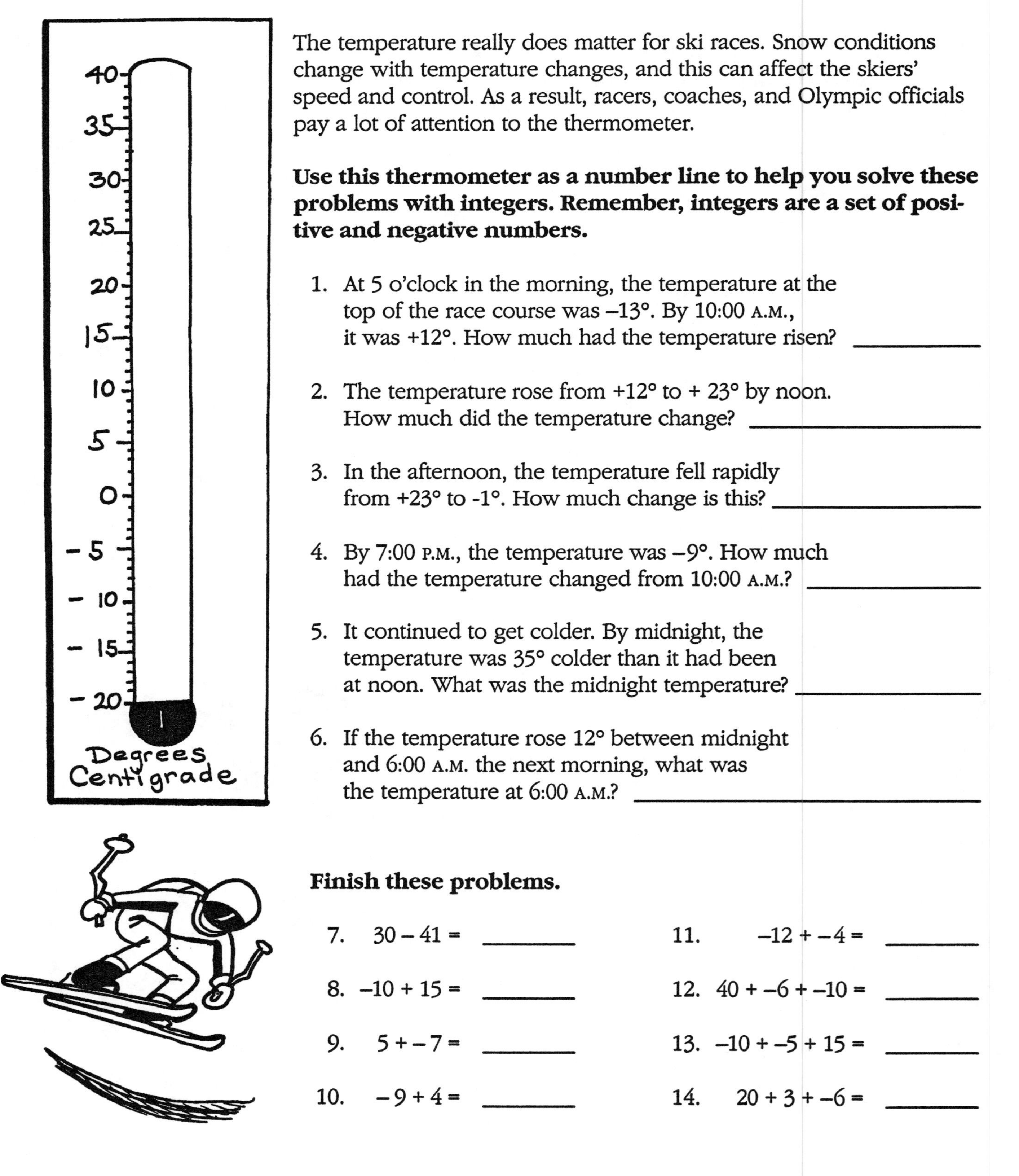

The temperature really does matter for ski races. Snow conditions change with temperature changes, and this can affect the skiers' speed and control. As a result, racers, coaches, and Olympic officials pay a lot of attention to the thermometer.

Use this thermometer as a number line to help you solve these problems with integers. Remember, integers are a set of positive and negative numbers.

1. At 5 o'clock in the morning, the temperature at the top of the race course was –13°. By 10:00 A.M., it was +12°. How much had the temperature risen? ________
2. The temperature rose from +12° to + 23° by noon. How much did the temperature change? ________
3. In the afternoon, the temperature fell rapidly from +23° to -1°. How much change is this? ________
4. By 7:00 P.M., the temperature was –9°. How much had the temperature changed from 10:00 A.M.? ________
5. It continued to get colder. By midnight, the temperature was 35° colder than it had been at noon. What was the midnight temperature? ________
6. If the temperature rose 12° between midnight and 6:00 A.M. the next morning, what was the temperature at 6:00 A.M.? ________

Finish these problems.

7. 30 – 41 = ________
8. –10 + 15 = ________
9. 5 + –7 = ________
10. –9 + 4 = ________
11. –12 + –4 = ________
12. 40 + –6 + –10 = ________
13. –10 + –5 + 15 = ________
14. 20 + 3 + –6 = ________

Name ______________________________

APPENDIX

Contents

GLOSSARY OF MATH TERMS

addend — a number added to another number.
In the number sentence 3 + 2 = 5, the addends are 3 and 2.

common factor — a whole number that is a factor of 2 or more numbers.
4 is a factor that is common to 4, 8, 12, and 16.

commutative property for addition and multiplication —
rules stating that the order of the addends or the factors has no effect on the answer.
5 + 7 = 7 + 5 and 6 x 8 = 8 x 6.

decimal — a number that uses a decimal point to show tenths, hundredths, thousandths, and other parts of a whole *(such as 0.2 or 0.02 or 2.005).*

decimal point — the dot in a decimal number.

denominator — the bottom number in a fraction. It tells how many in all.
In $\frac{4}{5}$, 5 is the denominator.

difference — the answer in a subtraction problem.

digit — the symbol 0, 1, 2, 3, 4, 5, 6, 7, 8, or 9.

equation — a math sentence that shows two things are equal.
20 x 4 = 80 is an equation.

equivalent fraction — fractions that have the same value.
$\frac{2}{4}$ and $\frac{1}{2}$ are equivalent fractions.

estimate — an answer that is not exact.

even number — a number that has 0, 2, 4, 6, or 8 in the ones' place.

fact family — a group of facts with the same numbers.
2 + 4 = 6, 4 + 2 = 6, 6 – 2 = 4, or 6 – 4 = 2 is a fact family.

factor — a number that is multiplied by another number.
3 and 4 are factors in the number sentence 3 x 4 = 12.

fraction — a number such as $\frac{1}{2}$, $\frac{3}{4}$, or $\frac{7}{8}$.

greatest common factor — the largest number that is a factor of 2 numbers.
8 is the greatest common factor of 16 and 24.

integer — any member of the set of positive or negative counting numbers and zero.
. . . –4, –3, –2, –1, 0, 1, 2, 3, 4 . . .

mixed number — a whole number plus a fraction, such as $1\frac{1}{2}$, or a whole number plus a decimal, such as 55.6.

negative integer — one of a set of counting numbers that is less than 0.

number sentence — a sentence that shows how numbers are related.
5 + 4 = 9 is a number sentence.

numerator — the top number in a fraction. It tells how many parts of the whole are being used or discussed.
In $\frac{2}{3}$, 2 is the numerator.

odd number — a number that has 1, 3, 5, 7, or 9 in the ones' place. It cannot be evenly divided by 2.
5, 767, 33, and 91 are odd numbers.

opposites property — rule that says that subtraction is the opposite of addition and division is the opposite of multiplication.
Since 6 + 8 = 14, then 14 – 6 = 8.
Since 7 x 5 = 35, then 35 ÷ 7 = 5.

place value — the value a digit has in a number, such as one, ten, hundred, thousand, or ten thousand.
In the number 437, 4 is in the hundreds place, so it has a value of four hundred.

positive integer — one of a set of counting numbers that is greater than 0.

product — the answer in multiplication.
In 6 x 7 = 42, the product is 42.

property of one — rule that states that a number multiplied by 1 has a product the same as the number.
5 x 1 = 5　　*600 x 1 = 600*

quotient — the answer in a division problem.
In 50 ÷ 10 = 5, the quotient is 5.

remainder — the number left over when a division problem is complete.
27 ÷ 5 = 5 with a remainder of 2.

sum — the answer in addition.
In 4 + 9 = 13, the sum is 13.

whole number — a member of the set of numbers 0, 1, 2, 3, 4, 5 . . .

zero — the word name for the digit 0, which means none.

zero property of addition and subtraction —
rule stating that adding zero to a number or subtracting zero from a number does not change the number.
60 + 0 = 60　　*60 – 0 = 60*

zero property of multiplication —
rule stating that any number multiplied by zero gives a product of 0.
55 x 0 = 0

COMPUTATION & NUMBERS
SKILLS TEST

Answer the questions. Each question is worth 1 point.

Write these numerals in words.

____ 1. 35,100

____ 2. 1,008

____ 3. 2,600,000

Write the numerals to match these words.

____________ 4. ten thousand, five hundred

____________ 5. three billion

____________ 6. fifty-five thousand, nine hundred, twenty

____________ 7. six hundred thousand, six hundred

____________ 8. eight thousand, thirty

Tell the place value of the <u>underlined</u> digit.

____________ 9. 5<u>3</u>62

____________ 10. 19<u>9</u>,527

____________ 11. 30<u>4</u>,111,111

____________ 12. 17,6<u>5</u>0

13. Number these numerals in order from smallest to largest.

_____ 500,005 _____ 25,500 _____ 3,005 _____ 5,505

____________________ 14. Round this number to the nearest ten: 5,976

____________________ 15. Round this number to the nearest thousand: 24,130

____________________ 16. Round this number to the nearest ten thousand: 278,166

____________________ 17. Round this number to the nearest hundred: 555

____________________ 18. Write the factors of 24.

____________________ 19. Write the factors of 49.

____________________ 20. Write the common factors of 12 and 32.

____________________ 21. Write the greatest common factor of 50 and 15.

Name

Skills Test

Solve these problems. Write the answer on the line.

__________ 22. $8\overline{)709}$

__________ 23. 5961 + 288

__________ 24. 710,621 − 25,009

__________ 25. 279 x 18

__________ 26. 648 ÷ 2

27. 5,500 ÷ 100 = __________

28. 65 x 100 = __________

29. 900 x 1000 = __________

30. 2,400 ÷ 10 = __________

31. 75 + 5 ÷ 2 − 20 = __________

32. 100 x 10 − 800 + 50 ÷ 50 = __________

Write the letter of the property used for each example below.

C = Commutative Property of Addition or Multiplication
Z = Zero Property for Addition or Subtraction
ZM = Zero Property for Multiplication
P = Property of 1

____ 33. 12 x 6 = 6 x 12

____ 34. 75 − 0 = 75

____ 35. 62 + 17 = 17 + 62

____ 36. 432 x 0 = 0

____ 37. 550 x 1 = 550

Write a fraction in each blank.

__________ 38. How many of the athletes are jumping?

__________ 39. How many of the athletes are riding something?

__________ 40. How many of the athletes are using some equipment?

Write a fraction or mixed numeral to match the words.

________ 41. seven twelfths

________ 42. twenty-two and eight ninths

Write a decimal numeral to match the words.

________ 43. fifty-five and five tenths

________ 44. three hundred forty-four thousandths

Name ____________________

Skills Test

45. Circle the largest fraction: $\frac{2}{3}$ $\frac{7}{8}$

46. Circle the largest fraction: $\frac{2}{12}$ $\frac{3}{9}$

47. Write the fractions in order from smallest to largest. $\frac{14}{16}$ $\frac{7}{12}$ $\frac{2}{3}$ $\frac{6}{8}$

____

48. Circle the fractions that are equivalent to $\frac{3}{4}$. $\frac{12}{16}$ $\frac{9}{12}$ $\frac{6}{12}$ $\frac{2}{3}$ $\frac{6}{8}$

49. Circle the fractions that are in lowest terms. $\frac{7}{9}$ $\frac{12}{15}$ $\frac{6}{12}$ $\frac{2}{3}$ $\frac{6}{8}$

__________ 50. Write this fraction in lowest terms: $\frac{18}{24}$

__________ 51. Write this fraction in lowest terms: $\frac{12}{16}$

__________ 52. Write this fraction as a mixed numeral: $\frac{50}{20}$

__________ 53. Write this fraction as a mixed numeral: $\frac{37}{5}$

__________ 54. Write this fraction as a mixed numeral: $\frac{44}{10}$

__________ 55. Write this mixed numeral as a fraction: $7\frac{1}{8}$

__________ 56. Write this mixed numeral as a fraction: $30\frac{5}{7}$

Choose the decimal numeral that matches the words.

__________ 57. seven and seven tenths — A. 0.05

__________ 58. five thousandths — B. 7.09

__________ 59. five hundredths — C. 0.0007

__________ 60. seven and nine hundredths — D. 0.005

__________ 61. seven ten thousandths — E. 7.7

__________ 62. seven and seven hundredths — F. 7.07

__________ 63. Round to the nearest thousandth: 0.48075

__________ 64. Round to the nearest ten thousandth: 1.86537

__________ 65. Round to the nearest hundredth: 1.6326

__________ 66. Round to the nearest tenth: 5.886

67. Write these decimals in order from smallest to largest. 0.03 3.003 0.3 3.3 33.33

____

Name ____________________

68. Circle the largest decimal numeral. 0.555 0.5 0.05 0.5005

Solve these problems.

69. $\frac{6}{7} \times \frac{2}{3} =$ ________
70. $\frac{7}{8} - \frac{2}{4} =$ ________
71. $\frac{6}{12} + \frac{2}{3} =$ ________
72. $\frac{7}{10} - \frac{4}{10} =$ ________
73. $\frac{3}{5} \times \frac{6}{9} =$ ________
74. $\frac{3}{4} \div \frac{2}{5} =$ ________
75. $\frac{5}{12} \div \frac{5}{6} =$ ________
76. $2\frac{1}{2} + 5\frac{1}{2} =$ ________
77. $1\frac{1}{2} \times 2\frac{1}{3} =$ ________
78. $\frac{12}{5} - \frac{4}{3} =$ ________
79. $16.5 - 4.2 + 2.1 =$ ________
80. $15.4 + 1.2 - 4.1 =$ ________

81. $ 200.40 − 15.80

82. $ 1600.99 + 743.86

83. $ 29.50 x 7

84. 3 ⟌ $75.00

________ 85. Change $\frac{3}{4}$ to a decimal.

________ 86. Write 72.3% as a decimal.

________ 87. Write 47.36 as a percent.

________ 88. Write 0.53 as a percent.

________ 89. Write 0.60 as a fraction.

________ 90. Change $\frac{2}{5}$ to a percent.

________ 91. Change 80% to a fraction.

92. Write these integers in order from smallest to largest: 5, −8, 3, 0, −2, 8, −5, 1

____

Solve these problems:

93. $-60 + 15 =$ ________
94. $13 - 7 + 4 + -8 =$ ________
95. $25 + 5 + -18 =$ ________

________ 96. Seven skaters need new skates. The skates cost $758.00 a pair. How much will the skates cost all together?

________ 97. Of the 42 boxers at the arena, $\frac{2}{3}$ had parents watching the matches. How many boxers is this?

________ 98. 10,700 athletes attended the Atlanta Olympics. About 7,700 fewer attended the Nagano Winter Games. How many athletes were in Nagano?

________ 99. The thirsty runners drank 24,375 gallons of water during the week. There were 8,125 athletes. On the average, how much did each one drink?

________ 100. 37,077 fans watched one basketball game. There were only 34,825 seats. How many fans had to stand?

Name ____

ANSWER KEY

Skills Test

1. thirty-five thousand, one hundred
2. one thousand, eight
3. two million, six hundred thousand
4. 10,500
5. 3,000,000,000
6. 55,920
7. 600,600
8. 8,030
9. hundreds
10. one thousands
11. one millions
12. tens
13. 4, 3, 1, 2
14. 5,980
15. 24,000
16. 280,000
17. 600
18. 1, 2, 3, 4, 6, 8, 12, 24
19. 1, 7, 49
20. 1, 2, 4
21. 5
22. 88 R5
23. 6,249
24. 685,612
25. 5,022
26. 324
27. 55
28. 6,500
29. 900,000
30. 240
31. 20
32. 5
33. C
34. Z
35. C
36. ZM
37. P
38. $^1/_5$
39. $^2/_5$
40. $^4/_5$
41. $^7/_{12}$
42. $22^8/_9$
43. 55.5
44. 0.344
45. $^7/_8$
46. $^3/_9$
47. $^7/_{12}$; $^2/_3$; $^6/_8$; $^{14}/_{16}$
48. $^{12}/_{16}$; $^9/_{12}$; $^6/_8$
49. $^7/_9$; $^2/_3$
50. $^3/_4$
51. $^3/_4$
52. $2^{10}/_{20}$ or $2^1/_2$
53. $7^2/_5$
54. $4^4/_{10}$ or $4^2/_5$
55. $^{57}/_8$
56. $^{215}/_7$
57. E
58. D
59. A
60. B
61. C
62. F
63. 0.481
64. 1.8654
65. 1.63
66. 5.9
67. 0.03; 0.3; 3.003; 3.3; 33.33
68. 0.555
69. $^4/_7$
70. $^3/_8$
71. $^{14}/_{12}$ or $1^2/_{12}$ or $1^1/_6$
72. $^3/_{10}$
73. $^{18}/_{45}$ or $^2/_5$
74. $^{15}/_8$ or $1^7/_8$
75. $^1/_2$
76. 8
77. $^{21}/_6$ or $3^3/_6$ or $3^1/_2$
78. $^{16}/_{15}$ or $1^1/_{15}$
79. 14.4
80. 12.5
81. $184.60
82. $2,344.85
83. $206.50
84. $25.00
85. 0.75
86. 0.723
87. 4,736%
88. 53%
89. $^6/_{10}$ or $^3/_5$
90. 40%
91. $^{80}/_{100}$ or $^4/_5$
92. −8, −5, −2, 0, 1, 3, 5, 8
93. −45
94. 2
95. 12
96. $5,306.00
97. 28
98. 3,000
99. 3 gal
100. 2,252

Skills Exercises

pages 10–11

1. ten thousand, seven hundred, fifty
2. one billion, six hundred million
3. two million
4. forty thousand
5. three thousand
6. forty million, one hundred thousand
7. one thousand, nine hundred, thirty-three
8. eighty-five thousand
9. eight thousand, sixty-two
10. three hundred fifty thousand
11. 35,000,000,000
12. 95,000
13. 4,099
14. $30,000,000
15. 10,563
16. $1,500,000,000
17. 29,228
18. 1,300,000
19. 200,000
20. 3,000
21. $394,000
22. 10,500

page 12

Down

1. 40,973
2. 151,600
3. 900,901
7. 71,800,003
8. 610,390
9. 450,009

Across

4. 300,050,008
5. 2,600,900
6. 207
10. 12,008,035
11. 8,351
12. 909
13. 300,300

page 13

1. Order of numbers in front of names: 11; 7; 8; 2; 6; 4; 9; 5; 3; 10; 1; 12
2. Order of numbers in list: 5; 4; 6; 2; 1; 9; 7; 3; 8

page 14

1. 11,000
2. 270
3. 2,000
4. 360
5. 900
6. 7,490
7. 15,000
8. 5,000
9. 800,400
10. 750,000
11. 1,000,000
12. 101,330
13. 61,000
14. 775,000
15. 10,990
16. 770
17. 98,900
18. 610
19. 600
20. 56,000
21. 900
22. 20,000
23. 600,000
24. 1,510

page 15

1. 7 tens
2. 8 hundreds
3. 8 tens
4. 5 thousands
5. 3 thousands
6. 0 hundreds
7. 5 ten thousands
8. 9 thousands
9. 1 ten
10. 9 ten thousands
11. 6 tens
12. 8 hundreds
13. 8 hundreds
14. 0 tens
15. 5 hundreds

page 16

1. 11,345
2. 9,974
3. 1,023
4. 11,111
5. 1,816
6. 10,970

page 17

1. 271
2. 217
3. 14
4. 42
5. 274
6. 35
7. 407
8. 5,297
9. 3,702
10. 581
11. 5,654
12. 7,275
13. 721,320
14. 12,138

page 18

1. 56
2. 312
3. 522
4. 940
5. 113
6. 111
7. 1,254
8. 189
9. 515
10. 873
11. 125,649
12. 11,038
13. 7,495
14. 2,062
15. 2,892
16. 7,358

page 19

1. 5,493
2. 1,339
3. Hungary
4. 1,377
5. Spain
6. Italy
7. 376
8. 266
9. 955
10. 7,308
11. 23
12. 101

page 20

Accept any four correct factors.

1. 1, 2, 3, 6, 9, 18
2. 1, 2, 5, 10
3. 1, 3, 5, 15
4. 1, 3, 5, 9, 15, 45
5. 1, 2, 3, 6
6. 1, 2, 3, 6, 9, 10, 15, 30, 45, 90
7. 1, 3, 7, 21
8. 1, 3, 11, 33
9. 1, 2, 3, 4, 6, 8, 12, 24
10. 1, 2, 4, 7, 8, 14, 28, 56
11. 1, 2, 4, 8, 16, 32
12. 1, 3, 9, 27
13. 1, 2, 3, 6, 7, 14, 21, 42
14. 1, 2, 4, 5, 8, 10, 20, 40
15. 1, 3, 13, 39

page 21

1. 2
2. 3
3. 11
4. 14
5. 15
6. 6
7. 4
8. 3
9. 10
10. 12
11. 3
12. 4
13. 2
14. 2
15. 2
16. 10
17. 5
18. 5
19. 4
20. 7
21. 4
22. 10
23. 5
24. 10
25. 2
26. 3
27. 3

The driver dragged a garden rake to stop the sled.

page 22

Check to see that students have colored picture correctly, according to answers given in color code.

page 23

1. 700
2. 300
3. 950
4. 200
5. 1,460
6. 1,000
7. 1,700
8. 200
9. 4,000
10. 1,300
11. 910
12. 409
13. 1,589
14. 5,791

page 24

1. 230 m
2. 2,300 m
3. 930 m
4. 111,000 m
5. 5,050 m
6. 888,800 m
7. 7,170,000 m
8. 80,480 m
9. 2,800 m
10. 250,000 m
11. 440 m
12. 44 m
13. 22 m
14. 100 m
15. 10 m
16. 33 m
17. 880 m
18. 10 m
19. 70 m
20. 6,107 m

page 25

1. Different answer is 201; all other answers are 200.
2. Different answer is 795; all other answers are 792.
3. Different answer is 96; all other answers are 144.
4. Different answer is 588; all other answers are 506.
5. Different answer is 630; all other answers are 670.

page 26

Hurdle problem: 100

1. 161
2. 6
3. 2,002
4. 888
5. 1
6. 202
7. 1,000
8. 0

page 27

1. divide; 1,089
2. multiply; $1,972,886.00
3. multiply; 15,600
4. add; 265.03 km
5. multiply; 2,000
6. divide; $22.90
7. multiply; $300,000,000
8. multiply and add; $3,944.00

page 28

1. commutative property of mult.
2. property of 1
3. zero property of mult.
4. commutative property of mult.
5. property of 1
6. commutative property of mult.
7. commutative property of addition
8. zero property of subtraction
9. opposites property of addition
10. zero property of addition
11. commutative property of mult.
12. zero property of mult.

page 29

1. $^2/_6$ or $^1/_3$
2. $^3/_6$ or $^1/_2$
3. $^3/_9$ or $^1/_3$
4. $^1/_9$
5. $^6/_9$ or $^2/_3$
6. $^5/_6$
7. $^2/_8$ or $^1/_4$
8. $^1/_8$
9. $^2/_6$ or $^1/_3$
10. $^4/_{12}$ or $^1/_3$
11. $^1/_6$
12. $^4/_6$ or $^2/_3$
13. $^4/_{12}$ or $^1/_3$
14. $^{11}/_{12}$
15. $^2/_6$ or $^1/_3$
16. $^2/_6$ or $^1/_3$
17. $^1/_4$
18. $^1/_6$
19. $^4/_{10}$ or $^2/_5$

page 30

1. K
2. AT
3. SPE
4. G
5. DS
6. I
7. E
8. N

Speed skating

1. N
2. L

Answer Key

3. OW
4. L
5. D
6. HI

Downhill

page 31

Correct fractions to color: 3, 4, 5, 6, 10, 11, 13, 16, 19, 20

1. $^3/_4$
2. $^1/_2$
7. $^4/_5$
8. $^1/_2$
9. $^1/_4$
12. $^3/_4$
14. $^1/_2$
15. $^6/_{11}$
17. $^4/_5$
18. $^2/_5$

page 32

1. $^2/_4$
2. $^5/_7$
3. both
4. $^1/_3$
5. $^1/_3$
6. $^5/_6$
7. $^7/_8$
8. both
9. $^2/_3$
10. $^{11}/_{12}$
11. $^5/_6$
12. both
13. $^1/_4$; $^2/_5$; $^1/_2$
14. $^3/_{18}$; $^2/_3$; $^5/_6$
15. $^2/_5$; $^5/_9$; $^6/_7$

page 33

Path follows these equations:

$^8/_{12} = ^2/_3$
$^2/_4 = ^5/_{10}$
$^6/_3 = ^8/_4$
$^8/_4 = ^{12}/_6$
$^{20}/_{25} = ^4/_5$
$^7/_{12} = ^{14}/_{24}$
$^0/_2 = ^0/_4$
$^2/_3 = ^4/_6$

page 34

1. C
2. B
3. B
4. A
5. C
6. B
7. A
8. A
9. B
10. C

page 35

1. $28^1/_2$
2. $9^3/_6$ or $9^1/_2$
3. $24^1/_4$
4. $23^1/_2$
5. $6^2/_5$
6. $10^5/_8$
7. $29^1/_3$
8. $20^3/_4$
9. $12^1/_4$
10. $27^1/_4$
11. $23^2/_3$
12. $15^3/_4$
13. $27^3/_6$ or $27^1/_2$
14. $10^9/_{12}$ or $10^3/_4$
15. $25^1/_4$
16. $4^2/_3$

page 36

a. $1^1/_4$
b. $2^1/_7$
c. $1^6/_7$
d. $1^1/_5$
e. $8^8/_{10}$ or $8^4/_5$
f. $3^1/_3$
g. $1^1/_2$
h. $1^4/_{16}$ or $1^1/_4$
i. $1^6/_{10}$ or $1^3/_5$
j. $1^4/_{11}$
k. $4^6/_8$ or $4^3/_4$
l. $4^2/_4$ or $4^1/_2$
m. $4^5/_{10}$ or $4^1/_2$
n. $1^3/_4$
o. $13^2/_6$ or $13^1/_3$
p. $5^2/_3$
q. $8^4/_6$ or $8^2/_3$
r. $1^2/_9$
s. $6^3/_5$
t. $5^1/_{20}$
u. $10^3/_6$ or $10^1/_2$
v. $1^1/_5$
w. $2^5/_7$
x. $1^1/_3$

page 37

See that fractions are circled with correct colors.

1. $^5/_6$
2. $^3/_{10}$
3. $^1/_{12}$
4. $^1/_8$
5. $^1/_2$
6. $^4/_5$
7. $^1/_6$
8. $^{13}/_{22}$
9. $^1/_3$
10. $^4/_9$
11. $^7/_9$
12. $^{19}/_{21}$
13. $^1/_2$
14. $^{19}/_{24}$

page 38

1. $^{19}/_{10}$ or $1^9/_{10}$
2. $^6/_9$ or $^2/_3$
3. $^7/_{13}$
4. $^4/_6$ or $^2/_3$
5. $^{15}/_{20}$ or $^3/_4$
6. $^6/_{11}$
7. $^5/_5$ or 1
8. $^{15}/_{16}$
9. $^{10}/_{25}$ or $^2/_5$
10. $^2/_{12}$ or $^1/_6$
11. $^{29}/_6$ or $4^5/_6$
12. $^{71}/_{100}$
13. $^{16}/_{30}$ or $^8/_{15}$
14. $^{25}/_{10}$ or $2^5/_{10}$ or $2^1/_2$

page 39

1. $11^6/_8$ or $11^3/_4$
2. $26^5/_8$
3. $17^2/_5$
4. $10^1/_4$
5. $11^4/_4$ or 12
6. $2^4/_5$
7. $10^7/_6$ or $11^1/_6$
8. $10^3/_6$ or $10^1/_2$
9. $^4/_5$
10. $10^1/_{11}$
11. $6^4/_9$
12. $19^4/_5$
13. $24^4/_5$
14. $14^2/_6$ or $14^1/_3$
15. $5^{10}/_{10}$ or 6

page 40

$3^4/_6$ or $3^2/_3$ pounds of potatoes
$5^6/_{12}$ or $5^1/_2$ quarts water
$1^1/_3$ onions
$5^5/_{12}$ cups broth
$3^1/_9$ carrots
$5^2/_3$ celery sticks
1 green pepper
$3^5/_9$ cups corn
$3^1/_6$ pounds mushrooms
$4^5/_6$ cups chicken
$2^2/_9$ t salt
$4^2/_9$ T herbs

page 41

1. $^{24}/_{28}$ or $^6/_7$
2. $^8/_7$ or $1^1/_7$
3. $^{27}/_{22}$ or $1^5/_{22}$
4. $^{10}/_3$ or $3^1/_3$
5. $^{20}/_{60}$ or $^1/_3$
6. $^{10}/_{36}$ or $^5/_{18}$
7. $^{42}/_{40}$ or $1^1/_{20}$
8. $^{100}/_{121}$
9. $^1/_4$
10. $^{36}/_5$ or $7^1/_5$
11. $^{15}/_{12}$ or $1^3/_{12}$ or $1^1/_4$
12. $^{32}/_{27}$ or $1^5/_{27}$
13. $^5/_{12}$
14. $^{12}/_{12}$ or 1
15. $^4/_{25}$

page 42

1. .0183
2. 1.083
3. .115
4. .55
5. 99.7
6. 13.4
7. 500.005
8. 9.78
9. .97
10. 10.83
11. 108.3
12. .234
13. 5.555
14. .978
15. 978.3
16. 11.501

page 43

Trick 1
7; 8; 4; 2; 5; 3; 1; 6
Trick 2
3; 6; 8; 7; 2; 5; 1; 4
Trick 3
3; 5; 6; 8; 2; 1; 4; 7
Trick 4
4; 1; 6; 5; 3; 8; 2; 7
Trick 5
7; 6; 8; 1; 3; 5; 2; 4
Trick 6
4; 5; 6; 2; 3; 1; 7; 8
Trick 7
7; 5; 1; 3; 8; 4; 2; 6

page 44

1. .1
2. .9
3. 10.1
4. .3
5. 1.6
6. 4.9
7. 2.2
8. .1
9. .1
10. 4.8
11. .18
12. 2.81
13. .78
14. .06
15. 6.00
16. 100.48
17. .94
18. 1.37
19. 4.60
20. 3.67
21. .469
22. 4.679
23. 7.090

24. .056
25. 41.523
26. .199
27. 5.011
28. .022
29. .7478
30. .1999
31. .7400
32. 15.0289
33. 1.1515
34. 4.3337
35. .5902

page 45

Karin	39.479	4th
Sofia	39.272	5th
Elena	37.901	11th
Kim	38.464	9th
Kerri	38.886	6th
Tatiana	39.928	1st
Nina	38.562	7th
Larissa	38.545	8th
Svetlana	39.738	2nd
Olga	37.063	12th
Kathy	39.482	3rd
Tamara	38.289	10th

gold—Tatiana
silver—Svetlana
bronze—Kathy

page 46

1. $3.00
2. $2.60
3. $945.00
4. $210.00
5. $120.00
6. $26.80
7. $360.00
8. approx $133
9. $2,085.00
10. $26.83
11. $143.00
12. $725.00

page 47

1. 0.78
2. 0.80
3. 0.74
4. 0.72
5. 0.77
6. 0.81
7. 0.75
8. 0.73
9. 0.71
10. 0.75

page 48

Earvin	90
Tony	90
Christian	90
Clyde	80
Alan	95
David	80
Rimas	83
Charles	94
Aramis	80
Karls	75
Gintaros	67
Patrick	78
Larry	83
John	86
Sergejus	83
Stojko	80
Chris	87
Scottie	93

1. Gintaros
2. Sergejus
3. Alan, Charles, and Scottie
4. Alan

page 49

$^4/_5$ — 80%
$^3/_{10}$ — 30%
$^3/_{12}$ — 25%
$^6/_{40}$ — 15%
$^{50}/_{100}$ — 50%
$^3/_4$ — 75%
$^{11}/_{50}$ — 22%
$^9/_{20}$ — 45%
$^2/_5$ — 40%
$^9/_{45}$ — 20%
$^3/_5$ — 60%
$^7/_{140}$ — 5%
$^2/_{20}$ — 10%
$^9/_{10}$ — 90%

page 50

1. Missing numbers for countries from top to bottom:
   101; 27; 26; 50; 9; 15; 35; 5; 8; 9; 11; 7; 7; 10; 17; 17; 15; 3; 15; 8
2. Bulgaria, Brazil, Great Britain, Belarus
3. Poland & Spain
4. China, Italy, Ukraine
5. France
6. Belarus
7. China
8. Germany
9. 17
10. Canada

page 51

1. Missing numbers for countries from top to bottom:
   26; 8; 8; 20; 5; 3; 3; 9; 4; 1; 2; 1; 4; 2; 3; 1; 3; 1; 2; 0; 0; 0
2. 183
3. 119
4. 64
5. Russia
6. Canada
7. Germany
8. Canada
9. Italy
10. Netherlands

page 52

1. 25°
2. 11°
3. 24°
4. 21°
5. –12°
6. 0°
7. –11
8. +5
9. –2
10. –5
11. –16
12. +24
13. 0
14. +17